THE LANGUAGE OF HUMAN ECOLOGY: A GENERAL SYSTEMS PERSPECTIVE

Robert J. Griffore
Michigan State University

Lillian A. Phenice
Michigan State University

KENDALL/HUNT PUBLISHING COMPANY
4050 Westmark Drive   Dubuque, Iowa 52002

Copyright © 2001 by Kendall/Hunt Publishing Company

ISBN 0-7872-8004-6

All rights reserved. No part of this publication may be reproduced, stored in a retrieval system, or transmitted, in any form or by any means, electronic, mechanical, photocopying, recording, or otherwise without the prior written permission of the copyright owner.

Printed in the United States of America
10 9 8 7 6 5 4 3 2 1

Contents

Preface

Chapter One
A NEW PARADIGM   1

    A Multitude of Perspectives   1
    Reductionism and Oversimplification   2
    Multifinality in the Human Sciences   3
    Defining Human Ecology   3
    The Scope of Human Ecology   5
    The Growth and Evolution of Human Ecology   9
    Needed: A Unifying Language and Conceptual Structure   10
    Exercises   13

Chapter Two
WHAT IS A SYSTEM?   17

    Types of Systems   17
    Systems Concepts and Processes   20
    Boundaries of a System   24
    Components of Systems   25
    Systems Languages in the Social Sciences   30
    Synthesis of General Systems and Human Ecology   31
    Figure 2.1   33
    Figure 2.2   34
    Exercises   35

Chapter Three
FAMILY ECOLOGY: A SYSTEMIC LENS IN VIEWING FAMILY LIFE   39

    Historical Overview 39
    The Family and General Systems Theory 41
    Structural Components of the Family 42
    Governing Components of the Family 44
    Dynamic Components of the Family 47
    Information Processing Components of the Family 50
    Interrelationship Components of the Family 52
    Life Process 54
    The Nature of Time in Families 55
    Dimensions of Space 56
    The Power of the Theory: Family as an Ecosystem 57
    Figure 3.1 59
    Exercises 61

Chapter Four
PSYCHOLOGICAL ECOLOGY   65

    Psychological Environments 65
    Origins of Bronfenbrenner's Model 68
    Bronfenbrenner's Ecological Systems 69
    Ecology and Developmental Outcomes 71
    Ecological Systems and Development 74
    The Individual in Bronfenbrenner's Model 75
    A Bioecological Model 75
    Applications: Risk and Asset Approaches 76
    Evolving Concepts of Structures 76
    Psychological Ecology and General Systems Theory 77
    Exercises 81

Chapter Five
URBAN AND SOCIAL ECOLOGY   85

    Foundations of Urban Ecology   85
    Building on Basic Concepts   86
    Developments in Urban Ecology   87
    Contemporary Urban Ecology   88
    Urban Life and Human Development   89
    Sustainable Urban Environments   90
    Urban Ecology in Terms of Human Ecology   90
    Urban Ecology and General Systems Theory   91
    Social Ecology   93
    Environmental Justice and Injustice   95
    Social Ecology and General Systems Theory   97
    Exercises   101

Chapter Six
HUMAN ECOLOGY IN THE POSTMODERN ERA   105

    A Brief Sketch of Postmodernism   105
    Postmodernism and Human Ecology   107
    The Individual in Postmodernism and Human Ecology   107
    Deep Ecology: A Construction of Human Ecology   108
    Deep Ecology: Hazards of Being Postmodern   115
    Deep Ecology and General Systems Theory   115
    A Matter of Social Construction   117
    Human Ecology and Science in the Postmodern Era   121
    Exercises   125

Ecological Exercises   129

Bibliography   149

Index   169

Preface

At the present time, human ecology is not represented by a single scientific approach. There are many diverse approaches, each seeming to represent what human ecology is, as well as the standards of scholarship in human ecology. Each human ecological perspective is valid with regard to what it seeks to accomplish. Yet there is a need for a comprehensive, holistic perspective.

Many of the human sciences have developed unifying ecological perspectives, theorems, models and laws. For example, the study of economics has interwoven concepts and constructs of basic systemic principles with the study of ecology, thus moving toward unifying principles in a field called economic ecology. Unifying principles in the study of psychology and ecology have led to a variety of approaches known as ecological psychology and psychological ecology.

While these ecological perspectives are useful, it is necessary to organize the dialogue about human ecology at a holistic level. This can be done by refocusing the language of human ecology on general systems concepts to forge a new holistic paradigm for human ecology. General systems theory provides a language for achieving a new synthesis among the human sciences to examine the relationship that exists between humans and their environments.

A general systems analysis of human-environment relationships provides a view of the whole web of interconnectedness and patterns of human-environment relationship. Patterns, processes, and dynamic webs of interrelatedness become visible through mapping ecological reality in human systems using a general systems theory language applicable to systems in general. General systems theory seeks to reveal the whole of reality in all it interconnectedness.

In Chapter 2 we discuss systems and examine concepts from general systems theory relative to human ecology. In subsequent chapters we examine selected ecological perspectives, focusing on concepts from general systems theory.

Included with each chapter in this book are several exercises. These exercises are based on Bloom's (1956) taxonomy for categorizing educational objectives. Chapter exercises are designed to measure a variety of intellectual tasks, such as knowledge, comprehension, applications, synthesis and evaluation.

At the end of the book are several ecological exercises. They are holistic and integrative, as well as practical and applied.

The authors wish to acknowledge Dr. Linda Nelson, Michigan State University Professor Emeritus, for her editorial assistance and insightful comments during the preparation of this manuscript.

Chapter One

A New Paradigm

The 21st century brings many potentials and opportunities, as well as many challenges, struggles, and possible catastrophic outcomes in the interactions of humans with their environments. The rapid expansion of technical and entrepreneurial knowledge, along with advances in knowledge of human development and interdependency of humans with environments are together creating a paradigm shift of vast proportions.

Tested scientific methods are being stretched beyond their capabilities. It is no longer a clear and universally accepted maxim that science will help humans out of these dilemmas. Once scientific inquiry could rely on accepted tools and techniques, such as experimental and quasi-experimental designs. Now qualitative designs and methods are increasingly accepted, and what was once regarded as rigorous has been replaced by various other perspectives on the nature of truth and reality. The force of these changes has opened the door for new initiatives in understanding the complex human-environment relationship.

Perhaps these changes have come just in time. Within the last half century, as human problems have been identified, observed, interpreted, and synthesized, it has become clear that the development of the human organism does not occur in isolation but within an interplay of interactions between the organism and its environment. The awareness of this interdependence is becoming ubiquitous, and the premises of interdependence are coming to be accepted as a holistic way of thinking, sometimes called a "new paradigm".

A MULTITUDE OF PERSPECTIVES

At present there is no single scientific approach to the study of human development that is accepted by the entire community of scientific scholars. It is generally agreed that scientists engage in searching for the "truth" in different proposed ways.

There are solid foundations in the human sciences on which to build our future understandings of people and environments, such as the work of Freud, Piaget,

Skinner, Bandura, and many others. Yet, much of this work reflects the defining paradigms of existing and established natural and social sciences. These perspectives are clearly valid with regard to what they seek to accomplish. They each describe a portion of the reality of the human condition and human problems. However, none of these perspectives is comprehensive and holistic.

Thus, in surveying the considerable territory of today's disciplines and paradigms in the human sciences, it seems clear that no single disciplinary approach is suitable for all situations or all problems. Human problems are most effectively solved when there is "goodness of fit" between the scientific approach, the evidence that it yields, and practical applications. While the various human science perspectives have led to valuable insights, they have also resulted in new and more puzzling questions.

As scholars examined human development from diverse disciplines and perspectives, very different conclusions emerged, often directly contradicting each other. This has led to confusing, and often contradictory, explanations and accounts of the meaning and facts of human development. While these contradictions have produced vigorous discussion and an evolution of ideas in the scientific community, it seems that there are limits to the utility of controversy. It is useful to move from debates about methods and approaches to action in the interest of humans and their environments.

REDUCTIONISM AND OVERSIMPLIFICATION

Over time, social science has forged ahead in an effort to create an acceptable niche for the study of humans. The problem is that often when one creates an acceptable niche, it is by oversimplifying reality. Thus, early human development science was based on narrowly defined relationships of subsystems within a more general human system. Investigators developed concepts and constructs that were about subsystems of human development. Hence there was very little effort to bring together and synthesize larger systems. Such narrowly defined investigations were applauded, as one could account for the variables under study, as well as manipulate the statistics to explain outcomes.

In essence the inability to synthesize all parts into an understanding and prediction of the behavior of larger systems was not thought to be critical to human survival. According to Odum (1994):

> "...it was not critical to survival before because humans were embedded in, and protected by, the smooth functioning of a giant biosphere of ecological networks. ...now we are capable of damaging our own basis for support if we make changes that we do not understand" (p.3).

MULTIFINALITY IN THE HUMAN SCIENCES

It is now becoming clear that in the study of human development, scientific knowledge is rapidly moving the scientific community toward basic unifying principles, some of which are found in various disciplines. Capra (1994) calls this a shift of paradigms within the sciences.

Presently it is as though there is rapid movement toward multifinality of the human sciences. Inquiry in the human sciences has become congested with a crush of overlapping models. Multiple final end points are based on established independent discoveries that occur within disciplines. This need not be the future of the human sciences. The science of ecology offers the scientific community a constellation of concepts and constructs for establishing theorems, models, and laws to guide empirical research at the person level, family level, the community level, and the global level.

Recognizing this potential, many of the human sciences have developed unifying ecological perspectives, theorems, models and laws. Ecology is being constantly discovered in the social sciences (Boyden, 1986; Micklin, 1984). For example, the study of economics has interwoven concepts and constructs of basic systemic principles with the study of ecology, thus moving toward unifying principles and establishing a field called *economic ecology*. Similarly, unifying principles in the study of psychology and ecology have led to an area of study called *ecological psychology*.

DEFINING HUMAN ECOLOGY

Ecology is the scientific study of relationships between living organisms and environments. Human ecology is the scientific study of humans interacting with their environments. The essence of all human ecology is person-environment relationships. Environments, broadly defined, incorporate the organic and inorganic world (Campbell, 1983). Humans interact with *natural environments, human constructed environments*, and *social environments*.

Natural Environments

Natural environments include animate and the inanimate environments - plants, animals, mountains, rivers, oceans - all living and nonliving things. For the individual, the earliest natural environment is the prenatal environment in which a functioning placenta takes away waste and delivers oxygen, food, drugs, toxins, and hormones from the mother. This natural environment is not necessarily pure. For example, the natural environment of the infant can contain toxic substances in breast milk, such as lead, mercury, dioxin, PCB, carbon monoxide, formaldehyde, and volatile organic compounds (Wargo, 1998).

Dimensions of the natural environment, such as climate and temperature, can have powerful consequences for people. Many people are adversely affected by dark, dreary climates. They may develop seasonal affective disorder (SAD), which responds positively to more light, allowing the pineal gland to produce more melatonin. Other forms of ambient environmental energy may also have diverse effects on behavior and development.

Human Constructed Environments

Human constructed environments include architecture, building size, building accessibility, relationships of rooms, ambient temperature and lighting, colors, odors, sounds, physical dimensions, and furnishings and equipment. Humans relate to constructed environments on the basis of these characteristics. Inaccessible rooms are generally unoccupied. Dark, cold, hot, or unsafe structures and rooms are avoided.

Characteristics of human constructed environments can have both short-term and long-term effects. For example, housing arrangements and neighborhood characteristics can affect children's affiliation with each other, the quality of their play behavior, social skill development, and perhaps participation in gangs.

Natural environments are constantly destroyed and replaced by human constructed environment. Rainforests are destroyed daily. Wetlands are constantly destroyed and replaced by shopping malls and office buildings, factories, sports facilities, residential developments, and service installations. Ecosystems that have been established over a very long period of time fall to heavy equipment in a moment.

Social Environments

The dimensions of social environments are both real and perceived. They are important for children and adults of all ages. Examples include recreational, work, and family environments. A family includes two or more individuals within a shared context called the social environment. Patterns of interaction among family members can be measured objectively and in many respects are evident to all family members. On the other hand, the social environment of the family for each member is a *microenvironment* - a socially constructed environment, the dimensions of which are quite uniquely constructed by each individual.

Culture is socially constructed. Individuals are influenced by their culture to create their own psychological constructions. Because each family member constructs his or her family to be a unique microenvironment, no two family members experience the same family. Siblings have very different experiences and, therefore, have very different developmental outcomes, although they develop within the same socially organized group defined as a family.

School classrooms are also socially constructed. Since no two students construct a school classroom in the same way, no two students have the same experiences or are affected by classroom experiences in the same way. A college classroom is another example. Students perceive and construct very different understandings and perspectives of a college classroom. Since their experiences are possibly extremely variable, they have very different views of the value of this experience.

There are few simple and clear lines between human constructed environments and social environments, nor clear distinctions between what is physically constructed and what is socially constructed. It is very difficult to dissociate human-made objects, such as buildings, roads, vehicles, containers, and plastics, from experiences and behavior in and with the objects, accept with arbitrary boundaries.

THE SCOPE OF HUMAN ECOLOGY

The scope and essence of human ecology are evolving. At the present time, we describe human ecology through ten conceptual statements.

Human ecology is a multidimensional science.

Human ecology is characterized by very significant multidimensional range, scope, and diversity of approaches, perspectives, and concepts. In this respect, human ecology is not unlike many of the social sciences. For example, in psychology, psychodynamic theories have unique vocabularies that are clearly different from behavioristic theories. Humanistic, existential, and cognitive developmental theories all resonate with different concepts, constructs, and processes. Similar phenomena may be observed in anthropology, political science, and sociology.

Human ecology has central core concepts.

At the core of human ecology, one finds concepts related to humans, concepts related to human environments, and concepts related to the interaction of humans and environments (Westney, Brabble, Saxton, & Holloman, 1986). As related to humans, basic concepts include human development, resources, and wellness. As related to environments, the core includes, as we have seen, characteristics of the natural, social, and human constructed environments. Interaction aspects of human ecology incorporate quality of energy, interactions and relationships of life forces that govern and regulate the interactions of persons with environments.

Human ecology seeks to describe and explain.

In some respects human ecology is like other sciences in its fundamental orientation toward description and explanation of interactions of human and environments. Beyond description and explanation, human ecology addresses itself to bringing about the conditions that promote human development and effective human functioning. Moreover, human ecology identifies a professionally oriented field that focuses on preparing professionals who can advance the causes of human betterment.

Human ecology is interdisciplinary and transdisciplinary.

Interdisciplinary science is a mode of inquiry quite different from single discipline scholarship (Klein, 1990; Kockelmans, 1979; McCall, 1990; Roy, 1990; Winthrop, 1964). Human ecology scholars tend to be interdisciplinary, working at the margins of their disciplines. They also tend to be transdisciplinary in studying humans interacting with their environments. They represent a view of the world that features holism of knowledge and ethics (Hayward, 1994). Human ecology is a science concerned with nature, with a particular focus on humans (Hawley, 1944).

Human ecology is holistic.

Human ecological thought is holistic (Hayward, 1994). In the holistic human ecological scientific paradigm, all events are related in complex ways. Causality is represented as complex, circular, and reciprocal. Any modification of any aspect of the natural, human constructed, or social system necessarily results in changes in all other systems. Thus human ecology acknowledges, and is predicated on, global holism and global responsibility. It incorporates important issues of the sciences and humanities.

An ecological education prepares students to be global citizens who understand and accept responsibility for the present and future consequences of their actions. Students are informed of the need to make ecologically sound decisions concerning human resources, human capital, and human diversity.

There is a scientific place for understanding particular phenomena that are only accessible through precise analysis. However, higher levels of ecological organization cannot be understood based on knowledge of lower levels of organization. Ecological understanding depends on examining a level of ecological organization about which one has interest - not a lower and less incorporative level of analysis. For example, one cannot understand social behavior by studying individuals outside of social environments or by studying the human genome.

Human ecology is a science of ecosystems.

Ecosystem is a fundamental concept in human ecology. An ecosystem is the basic unit of organism-environment interaction. It results from interaction of living and nonliving entities in a particular place. In an ecosystem, individuals come and go, while the community is maintained over time. Thus, ecology is not fundamentally about individuals alone or environments alone.

Human ecology focuses on that which is essential to human life: water, air, energy and land, production, consumption, and outcomes. Human ecology is about how a society achieves effective and sustainable patterns of human-environment interaction over time. The human ecosystem is the total biosphere, and human adaptation requires sustainability of the biosphere. The problem is that the biosphere is being transformed by individual and collective human actions at a pace that makes its sustainability questionable (Campbell, 1983).

In human ecological inquiry, the focus is on ecosystems as the fundamental unit of analysis. However, as one examines various areas of scholarship that are associated with human ecology, the focal importance of the concept of an ecosystem varies. For example, the concept of an ecosystem is present in cultural ecology, social ecology, and political ecology (Campbell, 1983). It is not explicitly evident in the philosophical foundations of deep ecology or in spiritual ecology (Merchant, 1992), which features deep relationships of humans with the earth and unified, holistic consciousness.

Human ecology involves many forms of inquiry.

Ecological study might examine ecological organization, distribution of individual within ecosystems, patterns of adaptation, and processes of regulation. In human ecology it is held that humans are part of the world – in constant interaction with environments. Thus, ecological method cannot be based on the premise of the separateness of humans from environments. The focus is on relationship of humans with environments (Brown, 1993).

Human ecology is about species.

In biological ecology two concepts are fundamental: species and niche. Species refers to all members of a group of living organisms. Niche is defined in terms of evolution, in a community, of genetic characteristics in relation to other species (Young & Broussard, 1986). In human ecological terms, a population exists in relation to other populations, and their respective roles, positions, or niches are articulated within a human context.

Framing the subject matter of human ecology in these terms makes it clear that there is potential for human ecology to evolve as a specific area of traditional ecological study, based on ecological concepts and principles. At the same time, human ecology focuses primarily on humans (Hawley, 1944; Stephan, 1970), and thus human ecology is often considered to be set apart from ecology in general. This view is regularly reinforced by accepting the individual or the environment as the focal point.

Human ecology is grounded in values.

Human ecology is supported by a structure of values. The value system of human ecology features the notion of humans living in harmony with the natural world and rejection of the premise that humans are in control of nature. Nature must be protected, while human potentials are realized.

Human ecology supports person-environment relationships in which human potentials may be most effectively realized. In the history of human ecology, as well as in the present, this refers to focus on home economics, home management, home-school relationships, community ecology, work and family issues, and a host of other person-environment issues, all of which can be influenced to bring about the most effective outcomes for quality of life. Human ecology emphasizes that individuals are active change agents of their own environments.

Human ecology is a force of convergence and synthesis.

Odum (1989), critical of reductionist ecological perspectives, described systems ecology as an integrative force in ecology. Young (1991) recognized that human ecology is connective, for it tends to bring together or connect disparate disciplines, perspectives, and ideas. While this tendency is reflected in human ecology, it is also a challenge. Representatives of disparate disciplines and quite different perspectives do not always come together easily or willingly. Human ecology has accepted a challenge that other transdisciplinary fields have avoided. Young (1991) observes that human ecology attempts to understand community. Perhaps this is based on the fact that community is a fundamental unit in biological ecology.

THE GROWTH AND EVOLUTION OF HUMAN ECOLOGY

Human ecology is grounded in biological ecology and based on concepts of ecology. Many of the foundations of human ecology reflect a solid grounding in ecological concepts (Tansley, 1939). Despite the clarity of the central core of the science of ecology, human ecology evolved as a diverse panorama of perspectives, held together more by name than by conceptual structure or methods.

Revisiting the Origins of Human Ecology

Since many of the founders of human ecology were social scientists, they often exercise considerable latitude in formulating their ecological perspectives (Bruhn, 1974). In some cases the use of basic ecological concepts is minimal, and in other cases concepts, such as organism, community, ecosystem, and equilibrium, have meanings quite different from their original meanings in biological ecology. In the last quarter of the 20th century it was not at all uncommon at meetings of the Society for Human Ecology, for example, to ask repeatedly the question "What is human ecology?", and to answer the question in very diverse ways.

In human ecology, as in all areas of scholarship, the questions posed, the modes of inquiry, and the answers obtained, all reflect contemporary society as it is socially constructed. Human ecology is also sensitive to structures of academic disciplines. Since disciplines such as psychology, sociology, and anthropology all have their own unique conceptual structures and methods, it is not surprising that when these disciplines become ecological, they tend to represent ecology in very different ways. Thus human ecology is shaped by ecological orientations within the various human and social sciences, and this is a source of considerable variation. Psychological ecology is quite different from sociological ecology or ecology associated with other fields. Thus the conceptual framework of human ecology is, in short, a construction from multiple perspectives.

Human Ecology and Biological Ecology

It has been suggested that the many conceptual frameworks of human ecology do not necessarily reflect basic ecological concepts and principles. The solution to this problem is not to treat human ecology simply as a branch of biological ecology.

Biological ecology is not, in itself, a sufficient foundation for human ecology (Bennett, 1976; Darling, 1955-1956; McIntosh, 1985; Orlove, 1980; Stewart, 1986; Visvader, 1986). Biological ecology has many limitations in addressing culturally based potentials of humans. Thus human ecology is more than a branch of biological ecology, and humans are more than biological organisms.

NEEDED: A UNIFYING LANGUAGE AND CONCEPTUAL STRUCTURE

What is needed is a language and conceptual structure capable of uniting diverse ecological perspectives. We suggest for this purpose the language and conceptual structure of general systems theory (GST). Further, we suggest that human ecology will be a more useful transdisciplinary science to the extent that it is linked to the mainstream of ecology based on the conceptual structure of general systems theory.

The use of a general systems framework for describing, observing, and understanding the complex interrelationships of human systems and environmental systems allows a description of the whole. It provides a common language among disciplines such as biology, psychology, economics, ecology, and other social sciences that seek to integrate human ecological concepts and perspectives.

Convergence

General systems theory, as a common conceptual foundation, moves toward achieving a meaningful convergence of various fields of ecology. The boundaries of disciplines are constantly shifting and permeable. Disciplines are in constant interplay, providing the opportunities for transdisciplinarians to address common issues, common perspectives, and common methods. Similar phenomena can be discussed in two or more disciplines. As these issues are mapped onto two different frameworks, there is a potential for a new synthesis to be created. As Koestler (1964) pointed out, a process of bisociation can take place to bridge diverse frames of reference.

As scholars from various disciplines consider events in collaborative frameworks, they are likely to find integrative, transdisciplinary, ecological common ground. They begin to think like human ecologists. They recognize the need to integrate, as Jungen (1986) observed, a variety of modes and ideals, including measurement tools, techniques, and data. General systems theory is a fundamental language for guiding this process of scientific convergence.

Such a convergence is natural. The isolation of disciplines is not a natural state of the present structure of knowledge. Disciplines became isolated due to the construction of idiosyncratic concepts, terms, and methods. Due to the invention of unique neologisms and concepts, disciplines lose contact with ecological ideas by straying from concepts grounded in general systems theory. By using concepts grounded in GST, disciplines can collaborate toward an ecological perspective.

Organizing the Dialogue at a Holistic Level

In order to advance, human ecologists can gain from purging their dialogue of preconceptions and language associated with particular disciplines. An ecological method, defined by general systems concepts, forges a new paradigm for holistic thinking. It supports a systemic vision of the phenomena of human developmental systems and environmental systems in dynamic activity or interrelationship.

General systems theory promotes a basic reorientation in scientific thinking. It is not necessarily committed to any set of concepts, but instead makes use of whatever concepts have the potential of contributing to a synthesis at a more general level (Bertalanffy, 1968). This is a language that can make it possible to achieve a new synthesis among the human sciences to examine the relationship that exists between humans and their environments.

A general systems analysis of information arising out of the phenomenon of human-environment relationship allows scientists to see the whole or the web of interconnectedness and thus patterns, which can serve as clues in understanding this relationship. One is reminded of the story in which one blind man tries to describe the elephant by touching and feeling the elephant's long protruding trunk, another by touching and feeling the elephant's stout legs, another by touching and feeling the elephant's big belly and so forth. Each reveals only a part of the whole, and therefore the final outcome is not the elephant as we know it.

Although human ecologists are interested in wholes, they can never entirely understand all that occurs at any one point in time. However, it is possible to perceive important patterns, processes, and dynamic webs of interrelatedness. This means keeping one's eye steadily fixed upon the information derived from the various studies of human development and, at the same time, not losing sight of the surrounding environmental system. Mapping ecological reality requires a network of relationships - a web. This is reminiscent of a concept from cognitive psychology, that in order for perception to occur there must be some cognitive structures into which perceptions can be mapped (Neisser, 1967).

General systems theory presents a dialectical perspective. A general systems goal is "the formulation and derivations of those principles which are valid for systems in general" (Bertalanffy, 1968:5). General systems theory seeks to uncover the core of reality, thus explaining more of the whole with fewer theoretical statements. This approach restores understanding of wholeness; it does not decompose reality into competing parts.

Exercises
Chapter 1 CONCEPTS

Chapter One incorporates many important concepts. List and define what you see as the 5 most important concepts in this chapter.

1.

2.

3.

4.

5.

Chapter 1 EXAMINING THE CONCEPTS

Choose the three most important concepts or ideas in Chapter One. In your own words, explain why you believe they are the three most important concepts or ideas. You are encouraged to compare and contrast terms as you explain your choice of the three most important concepts.

1.

2.

3.

Chapter 1 APPLYING THE CONCEPTS

Two important concepts in this chapter are *natural environment* and *social environment*. Choose two other important concepts, and for each of these two concepts, write an example to show how it can be applied. Give the page number where your application would fit in the chapter.

1.

2.

Chapter 1 SYNTHESIS

Write a final paragraph for Chapter One. In your paragraph, bring together the most salient thoughts in the chapter and draw conclusions. Be as thorough as possible and link together the main ideas and concepts.

Chapter Two

What is a System?

There are various ways to describe systems. Systems are defined as a collection of two or more units or parts that are bonded to each other forming a synthesis of a whole. Other definitions of a system mention the set of interrelated coherent behavior that acts as a unit. For example, *Webster's Third New International Dictionary* (Gove, 1964:2322) defines system as "a complex unity formed of many often diverse parts subject to a common plan or serving a common purpose." The essence of this definition is that component parts or units are bonded to each other by a relationship and the elements belonging to the system's environments, creating a whole new system (Allen, 1978).

Emerging from this relationship of interaction is a new property, emergence, which is conceptualized as making up the whole. The whole is a new synthesis of parts. Holism is more than adding component parts. The bonded components are greater than the sum of their parts. These wholes emerging from interactions of the component parts are called systems. The importance of any investigation lies in the wholeness of a system and other times on an analysis of the relating parts (Laszlo, 1972).

Systems analysis concepts and generalizations allow for a multitude of situations to be observed. This is critical for investigating a variety of systems including levels of complexity of systems.

TYPES OF SYSTEMS

Several concepts are used to differentiate the descriptive types of systems or the levels of its complexity: **open system, closed system, subsystem, suprasystem, ecosystem (one type of system), morphogenic system, morphostatic systems**. Figure 2.1 shows the relationships of types of systems.

Open and Closed Systems

The organizational characteristics of systems can be classified under more general and descriptive functional components of inflow of energy, information or matter. A system is *open* to the extent that its boundaries allow the flow of energy, information, and resources into the system. All living systems are considered open systems that allow energy and matter to flow in and out of the system. There is an exchange of energy and matter with the environment. The degree of openness is determined by the permeability of the boundaries. In human systems, a degree of openness of intra and extrasystemic boundaries is a necessary condition. If boundaries are too open and permeable, this could be like opening the floodgates and could result in unanticipated consequences.

If the boundaries are too impermeable, as is found in *closed systems*, rigidity can stunt adaptation and change. Closed systems do not have properties that allow energy and matter to flow in or out. All living systems must remain open for this exchange to exist. Certain systems may filter in or out more than others and therefore, the property of permeability exists. Degrees of semipermeability help to act as boundaries or filters for the system and help to sustain, maintain, or adapt the system to changes.

Subsystems and Suprasystems

Within limits, open systems are in a network with other systems, *subsystems* (smaller component systems), and *suprasytems* (environments of a system). The components of subsystems, and suprasystems have properties that are the same as the properties of the open system as a whole. Since the whole can be quite complex, one often constructs a model - a simplified version of an open system - and studies this model. Model building allows the human mind to understand a small part of the system, for example human ecosystems.

Ecosystem

An *ecosytem* is a diverse and complex network of many systems, subsystems, and suprasystems. The term *ecosystem* originated with A. G. Tansley (1939). He suggested that the fundamental conception is the whole system, including the organism complex and the whole complex of physical factors forming what is called the environment. No living organism exists in isolation. Organisms interact with one another and with the chemical and physical components of the nonliving environment. The basic unit of organism-environment interaction resulting from the complex interplay of living and nonliving elements in a given area is called an ecosystem. These ecosystems, so formed, are the basic units of nature and are of various kinds and sizes.

Melson (1980) states that an ecosystem approach considers organisms as they exist in nature and studies their interdependence with each other and with the environment. Adding to this, Smith (1972) emphasized the regular interdependence of organisms within their environment through exchange of energy flow and cycling of nutrients.

Morphogenic and Morphostatic Systems

Morphogenic systems are characterized by growth and adaptive responses supportive to change within the system. *Morphostatic* systems are characterized by stability and a correction of any deviation in response to change.

Controlled and Uncontrolled Systems

Systems have certain characteristics that can be used to classify them as either ***controlled systems*** or ***uncontrolled systems***. The difference between the two systems depends upon whether there is a preferred state or condition.

Controlled systems have goals, which determine or orient toward one preferred state over another (Kuhn, 1975). The preferred state is maintained within some range. Beyond the preferred state, a correction occurs through negative feedbacks. One example of a controlled system is a thermostat that is set a particular temperature in a house and maintains that temperature. The temperature of the human body is regulated in a similar fashion. If the body heats up beyond the normal range of human body temperature, there is a corrective, and the body cools down.

Norbert Wiener, father of cybernetics, (1948) called a controlled system a ***cybernetic system***. This type of system is known as a homeostasis system, meaning that a given level of homeostasis is maintained (Kuhn, 1975). Human social systems are mostly controlled systems. For example, if there is an uncontrolled state such as a spontaneous riot or unorganized demonstration by citizens, the police and the national guard are called to bring the crowd under control.

Uncontrolled systems have neither goals nor preferred states or orientations. A fundamental feature of the uncontrolled system is that a variable is not maintained within limits (Kuhn, 1975). In uncontrolled human systems, there is an acceptance of whatever state the individuals or groups happen to reach through their interaction with their system parts, and no preconception is formed as to what is hoped to be achieved. This describes most natural systems.

Humans have tried to control natural systems but have found that "Mother Nature" is not easily controlled. Humans have tried to control natural lakes, weather, glaciers, erosion just to name a few. Consequences of natural disasters, such as flooding of towns and farmlands, and erosion of ocean front properties resulting in the toppling of beautiful beach homes, are examples of trying to regulate uncontrolled systems

Pattern and Acting Systems

Human organizational systems occupy a complete spectrum of systems, from pure controlled to pure uncontrolled. However, a rule of thumb is that most all living things are controlled systems. Within human controlled social systems, there are two types of systems: *pattern* and *acting* systems. Pattern systems are logical, abstract, and analytical, such as found in philosophy, personality, maps, languages, or theory, such as economic theory. Acting systems are persons, and, at larger system levels, corporations or the U.S. economy. The difference between the pattern system and the acting system is between behaviors and the explanations of behaviors. Both are important but different, somewhat like mind and matter (Kuhn, 1975).

In summary, a basic assumption of general systems is that phenomena can be conceptualized as systems, which are sets of interacting parts with relationships among them. There are also uniformities in the behavior and functioning of systems about which we can generalize across all levels of systems. By integrating a general systems perspective to the study of human systems we can get to the heart of what is meant by stability, maintenance, and evolution or change of the system. Concepts of a general systems theory attempt to show how and why a system evolved in a particular way and provide the basis for a prescription regarding what is necessary for a system to survive.

SYSTEMS CONCEPTS AND PROCESSES

Inputs, Throughputs, and Outputs

Inputs are movement of matter, energy or information from an environmental source through the boundaries of a system. Resources flow from a source into a system. An input always modifies the system (Kuhn, 1974).

Throughputs are the transformation of matter, energy or information as it is processed from inputs to outputs. For example in the human body, as food is being digested, transformation takes place. Another example is the flow of information (input) of an impending hurricane to families who make decisions given the information (transformation) to seek shelter.

Outputs are movements of matter, energy, or information from within the system across its boundaries to the environment. This flow modifies the environment (Kuhn, 1974.). Outputs are outcomes of the transformation process. Using the examples from above, food being digested leaves the system as waste products or aid in the growth of an individual. A family makes a decision to seek shelter before the hurricane hits and weathers the storm.

Environment

Environment is all that surrounds and encompasses a system (Melson, 1980). An environment is not only external to the system, but also an internal context and should be viewed as part of the system as well. An example of an internal context would be the biochemical environment in the body.

Levels of Complexity of Systems

Systems vary in their levels of complexity. For example, Miller (1978) suggests that there are levels of systems within systems. In his development of general systems behavior theory this involves living systems, which are composed mainly of plants and animals. He suggests that there are levels of living systems with similarities in process and patterning. The levels are cell, organ, organism, group, organization, society, and supranational system. Systems at one level serve as parts of the next larger system. Fundamental component parts of systems apply to most systems at other levels of size and complexity.

The general principles of systems apply to the various levels of any system. For example, cellular systems form organ systems, which in turn form biological systems called organisms. The relationship of organism and components of environments form a system called *ecological systems*. Humanity together with natural environments are often called *environmental systems* (Odum, 1994). Humans in relationship to the natural environment, constructed environment, and social-cultural environment are called *human ecology* (Bubolz & Sontag, 1993).

Energy

Some form of energy is needed in any relationship. Therefore, a central emphasis of systems theory is what happens to this energy. Whether the energy is stored as a potential source or in constant flow, forms of energy are fundamental to all system processes. This includes energy stored in natural systems or stored in human resources, as well as in information flowing into the system as *input*. When the energy is processed by the system, it is often labeled as *throughput*, on its way to becoming an *output* of the system. The process of throughput or transformation of the energy through the system is governed by the rules of transformation. In living systems such as a family, family values, decision making, and rules help to transform the input.

Steady State and Dynamic Equilibrium

One of the most crucial concepts of systems is the relatively stable balance found between the two forces of inflows and outflows that make up the system. This balance is often referred to as *equilibrium* in most science and engineering fields. However, the concept of equilibrium as used within these disciplines often refers to states within closed systems. Within closed systems the storages and patterns of inflow and outflow become constant. Living systems are never closed. The term *dynamic equilibrium* is the concept of choice for living systems. This could be in the form of a cycle or in a relatively steady state. However, there are processes within living systems that can react to changes of input or output, which may influence the steady state nature of the system at a particular point in time. This then leads to another level of steady state or dynamic equilibrium. There is in systems the capacity to monitor or regulate progress toward set goals. This capacity is called feedback.

Feedback

In human systems, constant exchanges are going on in nature. Exchanges of energy, information, matter, and resources make for an open system. Let us suppose a system can be described as a bond between A and B. Because A is bonded to B, if A varies directly with a change in B, and B varies inversely with a change in A, there is movement toward equilibrium. This is known as *negative feedback*, or a corrective change (Kuhn, 1975). Negative feedback does not mean it is bad or mean.

Likewise *positive feedback* does not mean good or necessarily desirable. Positive feedback means the action is not opposite or corrective but in the same direction. Instead of stopping or slowing down the action, there is an acceleration in whatever direction has been set (Kuhn, 1975).

Processes of feedback operate at different **levels of complexity** (Broderick & Smith, 1979).

Level 1 feedback is *simple feedback*. Simple feedback is a circular process. An output is subsequently processed as input. In some situations, there is an amplification of the original input and spiraling. This is called a positive feedback loop. There is a cumulative effect of feedback processes in amplifying the deviations of a system in a particular direction away from a preexisting goal (Maruyama, 1963). This requires openness in a system. If, however, the result is dampening or a reduction of the original input, it is called a negative feedback loop. Negative feedback requires a degree of closure of a system to maintain stability, a homeostatic system.

Level 2 feedback is known as *cybernetic control*. There is some standard, criterion, or particular value by which the system is monitored. The output from the system is monitored and compared to the set criterion. Adjustments or corrections are made to any deviations from the set criterion. In human living systems these standards, policies, or goals may be called meta-rules. These rules are often based on the importance of values held by humans in that system.

Level 3 feedback involves *morphogenesis* of a system. When level 2 feedback (output) is not measured at a value within its range of corrective responses, an evaluation is necessary to change the meta-rules. The living system can move toward the goal.

Level 4 feedback leads to system *conversion* or *reorientation*. Here the goals of the system are changed, and there is a new goal.

Feedback is a critical process in system functioning. It regulates a system's capability to (a) monitor its own progress toward a set standard or goal, (b) correct and elaborate its response, and (c) even to change its goals if necessary. The state of a dynamic system depends upon the complexity of its feedback structure, otherwise without these capacities to monitor itself, the system becomes static (Broderick & Smith, 1979).

Human populations and social organizations are in constant state of flux. Because of the complexity of the processes involved, linear causality loops are rare in human systems, and reciprocal causality loops are more common. Negative feedback maintains a homeostatic system in which a high degree of closure of a system is required, and this is not found in many living human systems. Because human systems are not uniform and not homogeneous, positive feedback that amplifies the human system toward change is required.

BOUNDARIES OF A SYSTEM

Every system has a boundary. A boundary is a kind of division of what is inside from what is outside. It is the area of separation between systems or subsystems. Boundaries of a system are barriers to flows of matter, energy, and information as they move in and out of a system (Kuhn, 1975). In order to overcome barriers, more work must be expended to move matter-energy or information, except at points where some of the flow is permitted to cross into or out the system. Amounts of energy, information, and matter that flow may vary over time. Boundaries also help to maintain a steady state differential between the interior of the system and its environment.

Physical or Spatial Boundaries

Physical or spatial boundaries are the easiest to visualize or describe. For example, the surface of your skin creates a physiological boundary that filters out the sun's rays, which could destroy underlying tissue cells. The walls of a house surround and protect a family from the weather, intruders, and unfriendly interruptions. The State of Michigan is bounded by lakes and invisible surveyors' lines. Physiological boundaries could include the body membrane, the skin, and the hair, whereas psychological boundaries determine what is me and not me, often indicated by feelings and thoughts.

Functional Boundaries

Functional boundaries are often described by the functions of the systems, such as roles occupied by individuals in families or society. Mother works outside the family and brings in income to her family. Therefore, mother's provider role sets the income earning system boundary of this family. Her children are outside this income earning system, but are definitely part of the income using system. Psychological states, such as one's identity, memories, and one's defense mechanisms, are also functional boundaries, in the individual's personal system (Kuhn, 1974).

Analytical Boundaries

Analytical boundaries are often similar to functional boundaries. These boundaries are set for purposes of analyzing processes, interactions, or structural complexities of a system, for example the hierarchy of a system (Kuhn, 1975). For example, what system will be the focal point? Will it be the particles, atoms, molecules, cells, tissues, organs, organism, community, or society level? In the study of analyzing human processes and interactions, the interest could be the boundaries of the mind such as the systems of opinions and judgments, or decision making and actions (Hartmann, 1991).

Who Determines the Boundaries?

Boundaries are often set by the participant, observer, the culture, the investigator, or researcher. These boundaries are often determined and influenced by the state of knowledge available at a given time in history, or by the investigator interested in the particular system. The observer determines whether the focal point will be the individual, family, corporation, church, or nation. *Interface* is the common or shared boundary of systems.

The argument is not whether something is a system or not, but whether the perception of the observer is congruent with viewing it as a system, and at what level. Every system encloses subsystems. At the same time, suprasystems enclose systems. Thus there are many suprasystems, systems, and subsystems, down to the level of the quark, particle, or wave. In order to choose boundaries for analysis of a system, it is suggested that one models a system one size larger than the system under study.

COMPONENTS OF SYSTEMS

"A component is any interacting element in an acting system" (Kuhn, 1974:20).

Structural Components of Systems

Some components of systems are structural. Terms used to describe the composition or structural effects of a system include: **structure, hierarchy, and complexity**.

The *structure* of a system is the pattern of organization of its subsystems and functions. For example, in human systems a person's role is a patterned system and, therefore, presents the structural effects. The ascribed and prescribed human roles in organizations are a set of identified system states and define the actions of the subsystem. The person occupying a specified role is the acting system who performs behavior specified in the role. In human social systems almost all demographic characteristics, such as age, gender, income, levels educational attainment, number of subsystems within the system as well as personal characteristics of individuals, have effects on the structural components of the system.

Hierarchy in human systems consists of any relations between systems in which one is considered either a subsystem (a component of a larger system) or a suprasystem (the larger system in which a system is a component) (Kuhn, 1975). In human group relationships these are often determined by rank and order such a ***positions*** and ***status***, which suggests a subsystem or a suprasystem relationship.

Social class status and mobility such as in educational attainment, income, and occupation are often used to classify and determine the order in society. Facts of energy flow and distribution indicate that true power of individuals or groups lie in the useful potential energies that flow under their control. The assumption often made is that those with higher social class status or rank or order have more potential power and control over the flow of the concentrated energy, such as information, and material and human resources. In human families, the number of roles and the distribution of power within the family together forms a hierarchy and, therefore, the structural components of the family.

Complexity of a system is determined by the number of intricate parts. For example, examine a simple organization of two individuals' network of interactions and energy flows versus four individuals in the system. In either case, all individuals draw upon the same energy source. Through their actions of energy transfer and response to each other's information, they exert influences on each other. The pathways and expressions of these interactions are the same kind as flows found in energy transfers.

Parkinson's law states the general effect that efficiency diminishes as the number of units in the organization increases. If this is the case with social systems, then, as organizations include more individuals or social groups, the number of possible forward and return pathways rises rapidly between individuals and groups (Odum, 1994: 214). The coordinated functions involve a more complex network of pathways of materials, energy, and processes (Odum, 1994: 213).

In human social systems one could interpret this to mean that there may be a need to coordinate these flows and pathways to enable individuals to survive and thrive. At the same time, the larger the system's energy support, the more energy it has for the organization (Odum, 1994:216). This phenomenon is found in instances where an individual has a large supportive network of people to rely on in times of crisis. Here the social support group may enable the individual to achieve the goal.

Dynamic Components of Systems

Dynamic components reflect the flows of energy, matter, and information that are internal and nondisruptive to the system. These components include *internal flow processes*, *negative entropy (order)*, *energy*, and *matter*.

In open systems, internal flows function through differentiation and coordination of components. Negative entropy or order is maintained by the flow of matter, energy, or information which is reinforced by feedbacks of the system. Often this process is called throughput or the transformation of the input of matter, energy, or information within the system.

There are two very important energy laws. The first is that energy is neither created nor destroyed in a system. The second law is that heat is released from all storages and all processes. Therefore, as one thinks of potential energy, it is important to remember that potential energy is capable of driving a process with energy by transforming it from one form to another, but in this transformation heat is released. Heat is potential energy that is lost. For example, an energy transformation from one kind to another is work. When doing work, some of the potential energy is lost as heat.

Many forms of energy are involved in many different processes. In terms of the physics of energy, examples of forms include sound waves, sunlight, and chemicals which react. For example, solar energy is found as direct radiant energy from the sun. Indirect forms of solar energy include wind, falling and flowing water, and biomass in which solar energy is converted into chemical energy stored in the chemical bonds of organic compounds in trees and plants. In most types of work, one type of energy is being transformed from one form to another (Odum, 1994), and heat is released.

Kinetic energy is the energy of movement. All matter contains some storage of potential energy, therefore, all flows of matter into a system have a component of energy.

Some energy sources are *renewable* and can be used without depletion when carefully managed. Most of these sources come directly or indirectly from the sun as direct solar, wind, biomass or hydroelectric power. However, not all renewable energy exists without environmental consequences, for example, hydropower.

Nonrenewable resources such as copper, aluminum, coal, and oil, are found in a fixed amount in the earth's crust and have the potential for renewal only by geological, physical and chemical processes taking place over hundreds of millions of years. A non renewable resource is "one that is limited or depleted to such a degree that its recovery is cost prohibitive" (Naar, 1990).

Governing Components of Systems

Governing components are involved in the regulation and control of systems. These components include *control, cybernetics, feedback, homeostasis,* and *equilibrium.* More common terms serving similar functions may be found in human organizational systems, such as *rules, purposes, and goals.*

Dynamic equilibrium occurs when components or variables of a system continue to move or change, but at least one variable remains in a specified range. A balance is maintained. Examples of this can be seen in the regulation of dams where the constant height of the water equals the outflow of the water. "Homeostasis is the condition in which a controlled system maintains a steady state equilibrium of at least one or more system variables" (Kuhn, 1974:28).

Information Processing Components of Systems

Information processing aspects of a system are usually found at points of energy transformations and at intersections. These components include: *information, channel, memory, communication,* and *learning.*

Human systems are viewed by some social scientists as essentially forms of communication and control systems. It is believed that everything a person does conveys some kind of message. Therefore, it may be said that all human behavior is communication, and that individuals cannot avoid communicating some information in some way.

For this chapter, communication is defined as any transfer of information between systems or any movement of anything between systems which can be analyzed for its informational content. For example, a sign, a symbol, a referent, a code, a message, a nonlinguistic pantomime, or exaggerated gesture, as well as one's perception of the event or act embodies the informational content.

Thus, culture can be defined as a communicational phenomenon. According to Odum (1994) flows of information involve large concentrations of energy and have a high ratio of transformation. For example, Odum (1994) suggests that flows of books, genes, television, computer programs, art, and religious communication are all examples of information. Feedback from energy transformations allows a system to learn.

Interrelationship Components of Systems

Interrelationship components are concepts that take into account associations between parts of the system. Included are concepts such as *holism, interdependence, interaction, and independence.*

Holism refers to the system as a whole. Thus, to understand a system holistically one moves beyond examining a system solely in terms of its particular functional parts. An understanding of holism involves examining the uniqueness and multiplicity of components that make up a system. It also involves examining the webs of *interaction* in systems, in which the flow of information, matter, or energy creates complex reciprocal changes among system parts. There is *mutual interaction* when a change induced in one part of a system also changes another part (Kuhn, 1974).

Disruptive Components of Systems

Disruptive components reflect the disruptive forces that affect systems, including concepts such as *conflict, entropy (disorder), strain, stress,* and *threat.*

Increases in chaos or randomness and decreases of creativity accompany what is called *entropy* of a system or disorder within the system.

Life Process Components of Systems

Life process components are terms that reflect the nature of living systems. *Adaptation* is a dynamic process that is evidenced by different characteristics at different times within different contexts. In order for systems to adapt, system boundaries must be flexible, allowing change to occur. According to Kuhn (1974), in human systems, adaptation is viewed as behavior that changes the relations of the system to its environment, whether by changing itself, the environment, or both. Successful adaptation increases the probability of achieving some goal. Concepts related to the adaptation process may include processes of *change, evolution, growth, reproduction, decay,* or *termination.*

Changes such as marriage, births, maturation, children leaving home and creating an empty nest are all examples of structural changes within a family, but not necessarily changes for the society. What is considered change at one system level may not necessarily be an alteration of a structure at another level.

Decay is seen as a break up of a higher level of a system in which certain subsystems still remain. For example, if a family member dies, the original social system does not exist, but the remaining members of the family who were its subsystems still go on. One can say that when an individual can no longer function without a life support system, or when marriages end in divorce, or when businesses fail, these are examples of decay of the social system at that level.

SYSTEMS LANGUAGES IN THE SOCIAL SCIENCES

There are many systems languages arising from the holistic trend of both social and ecological sciences. These systems languages are not always clear and comparable. These languages tend to be clear where there are measurable procedures. Where definitions are vague, there is a tendency for confusion.

General systems thinking is clear and precise. The use of general systems language allows scholars in various fields to communicate across boundaries using a common vocabulary with shared meanings. The use of precise terminology and general systems language will facilitate thinking and communication among the various disciplines interested in human ecology. The seminal book *The Logic of Social Systems* by Kuhn (1974) clearly sets a foundation in integrating general systems thinking into the approach of studying the developing human environment relationship.

Because verbal language has its limitations, sometimes it is easier to use nonverbal or symbolic forms of expression. This is especially true of systems language Various groups such as engineers or biologists use diagrams and line drawings to express their understanding of systems under their scrutiny. This method is especially useful in understanding and thinking about systems. By diagraming the parts of a system, an inventory of the patterns of relationships can be found.

For example, energy circuit language (Odum, 1994) is a useful means to explain the flows of energy, information and matter within any living and nonliving system. The use of energy circuit language has a definite place in human ecological systems. Instead of reinventing a completely new symbolic language in the social sciences, modifying the use of general systems language can be an effective tool for communicating with those in different disciplines.

The human system is defined in terms of interrelationships among its parts. Many concepts have been used to reflect system processes, patterns, and functioning. In order to have a common language among all systems sciences, it is best to use concepts that are widely understood and to describe concepts in terms that can be understood by social scientists and/or physical scientists.

Those interested in understanding the relationships of systems across disciplines can thus communicate with each other. This facilitates transdisciplinary research and practice in human ecology. Common systems concepts provide the basis for a natural progression into a general systems framework. These commonly used concepts are essential to systems analysis and to an understanding of human systems characteristics of stability, adaptation, maintenance, and evolution. Using general systems terms and concepts allows for a precise description of human ecosystems, so that they can be investigated with minimal ambiguity and confusion. Concepts and principles from general systems theory that define the foundation of human ecosystems are as follows.

At its core, human ecology consists of common concepts and language. Some of these concepts have been developed, while others have yet to be considered. Human ecology will benefit from a systematic emphasis on common unifying principles and concepts. General systems theory is the fundamental foundation of human ecology. It is a force that can organize and provide a way to define, shape, and evolve contemporary human ecological approaches. In addition, general systems theory can provide, for human ecology, a language with which to communicate with scientists from other scientific disciplines and interdisciplinary perspectives.

SYNTHESIS OF GENERAL SYSTEMS AND HUMAN ECOLOGY

In the new human ecological synthesis, patterns of explanations - not generalizations or theories - emerge. One can take three different levels of perspectives to understand patterns: *paradigm level*, *regime level*, and *process level*. Depending on one's level of analysis, one tend to choose particular methods and to reveal particular levels of understanding. Patterns emerge within these three levels of categorization to provide a dynamic view of human development (see Figure 2.2).

Paradigm Level

The *paradigm* level is the highest level of analysis, focusing on patterns that provide a general world view or image. As an example, a family's paradigm is an image or a general model of what a family is, can be, or ought to be, thus serving as a point of reference for the family regime. The paradigm level is examined by looking at *suprasystems*. At this level, at any particular point in time, one often takes the perspective of a *functionalist*, studying human development and its role in the society. When one examines change and dynamics, the systems property that is unique to this level is *emergence*. Emergence refers to the development of a higher level system (Kuhn, 1974).

Process Level

The lowest level of analysis is that of *process,* which, in human development, can be thought of as behaviors and interactions. The process level is examined by looking at *subsystems.* At this level, at any particular point in time, one often takes the perspective of a *reductionist,* studying interpersonal processes in families. When one examines change and dynamics, the systems property that is unique to this level is decay. *Decay* is the breakdown of systems so that only components exist (Kuhn, 1974).

Regime Level

There is a level of analysis between the paradigm level and the process level. This is the level of *regime,* in which the organization, structure or regulating mechanism patterns are identified. This level of analysis is associated with the study of *systems.* A family's regime is that set of mechanisms by which collective patterns in process are regulated. To understand the whole system at a particular point in time, from this level one typically looks upward toward paradigms and suprasystems, or downward toward process and subsystems (holism). From a dynamic or developmental perspective, looking upwards one sees emergence. Looking downwards one sees decay (Kuhn, 1974).

Levels and Systems Thinking

Those who use a systems level of analysis are *holistic* in their approach, viewing human development within a context of systems. When all the levels of analysis are included, the patterns that derive are *dynamic,* which means that there are patterns of *emergence, continuation,* and *decay* in development. When one takes a dynamic developmental viewpoint, the three levels of suprasystem, system, and subsystem are all included. Subsystems exist because of exchanges occurring between the various levels of systems.

In many studies of human development, it is common to identify one of these levels for analysis as the subject of a static cross sectional view of the phenomenon that is being studied. Developmental psychology, in which one often focuses on *subsystem* levels of human development, may be considered reductionist, unlike the functionalist looks for pattern systems within a model, an image, or world view. There is a place for reductionists who may work within the boundaries of laboratory controls. However, reductionist approaches views need to be integrated, and a transactional approach is therefore imperative.

Figure 2.1

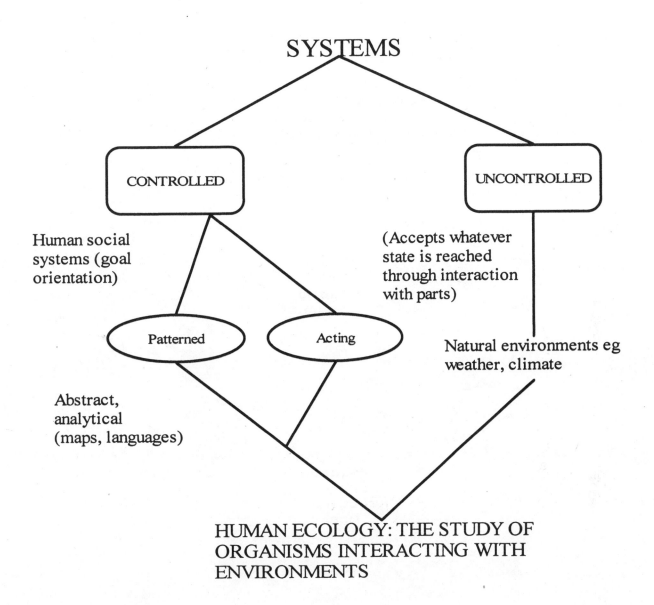

Figure 2.2

```
┌─────────────────┐
│                 │
│    PARADIGM     │
│   SUPRASYSTEM   │
│   FUNCTIONALIST │
│    EMERGENCE    │
│                 │
└─────────────────┘

┌─────────────────┐
│                 │
│     REGIME      │
│     SYSTEM      │
│    HOLISTIC     │
│   CONTINUANCE   │
│                 │
└─────────────────┘

┌─────────────────┐
│                 │
│    PROCESS      │
│   SUBSYSTEM     │
│  REDUCTIONIST   │
│     DECAY       │
│                 │
└─────────────────┘
```

THREE LEVELS

Adapted from Kuhn (1974).

Exercises
Chapter 2 CONCEPTS

This chapter about systems includes many important concepts. List and define what you see as the 5 most important concepts in this chapter.

1.

2.

3.

4.

5.

Chapter 2 EXAMINING THE CONCEPTS

From this chapter on systems, choose the three most important concepts or ideas, as you see them. Why do you believe they are the three most important concepts or ideas? Please compare and contrast terms as you explain your selection of the three most important concepts.

Chapter 2 APPLYING THE CONCEPTS

Two important concepts in this chapter are suprasystem and subsystem. Choose two other important concepts from this chapter. For each of the two concepts you choose, write an example to show how it can be applied. Give the page number where your application would fit in the chapter.

1.

2.

Chapter 2 SYNTHESIS

This chapter has focused on systems. Now that you have read it, compose a final paragraph in which you bring together ,show the relationships among, and draw conclusions about, the important ideas in this chapter.

Chapter Three

Family Ecology: A Systemic Lens in Viewing Family Life

Ecology is the science that studies the interactions between living organisms and their environment. The environment has organisms, water, chemical cycles, air, other people, machines, cities, forests, oceans as well as societies. And connecting them all are flows of energy.

An ecosystem is the basic unit of organism-environment interactions resulting from a dynamic interdependent web of relationships of living and nonliving elements in a given area.

Human ecology is the study of ecosystems, the fundamental interdependence of all phenomena and the embeddedness of individuals, families, and societies in the cyclical process of nature (Capra, 1994).

Family ecology is the study of family ecosystems which is the totality of relationships that define the family-environment system as an integrated whole.

HISTORICAL OVERVIEW

In 1892, Ellen Swallow Richards, the first American women scientist, saw the critical importance of the home and the family in the emerging "science of living"(Clarke, 1973). As a social reformist and a chemist with a degree from the Massachusetts Institute of Technology, she saw the importance of understanding the relationship of human health and behavior and the quality of the environment to improve the quality of people's lives. Richards saw that the environment was being challenged by increased manufacturing of goods and technology. She believed that balance was necessary so that people can have some control over their lives and environment. They needed knowledge from many sources, together with a structure for that knowledge and its applications. This was the underlying basis for the science of oekology that Ellen Richards announced to the world in 1892; "...let *oekology* be henceforth the science of (our) normal lives . . . " (Bubolz & Sontag, 1993:420).

She and others met at Lake Placid Club in New York State to discuss the founding of "home oekology." This was to be a science of the family and its relationship with the environment. The primary concern was to understand this relationship and to use the information to improve people's lives. There were insurmountable obstacles to Richards's oekology being accepted in the scientific world at that time. As discussed by Bubolz and Sontag (1993), during the late nineteenth and early twentieth century, most of the scientists were men, and most of Richards' followers were women. At that time, the accepted science of oekology included the relationships of plants and animals and their environments and did not include the organism called humans. The sciences of that time were discipline based, and what was being proposed as human oekology was a controversial interdisciplinary science.

From these discussions at Lake Placid, it became clear that Home Economics was the name preferred by this body. However, an ecological perspective was implicit in the 1902 American Home Economics Association's definition of this newly founded science (Brown, 1993). The approach was grounded in an interdisciplinary science which included the natural sciences, social sciences, as well as the arts and humanities, with an emphasis on application to everyday life. There have been significant contributions to this interdisciplinary study of the family in relationship with the environment.

For the purposes of research, practice, and policy development, this perspective of the family assumes that phenomena must be examined in their wholeness of interaction and interdependence with the environments. As Paolucci, Hall and Axinn (1977), write:

> " A family ecological perspective offers a holistic approach to the study of the family. It focuses on the family and those environments that affect it and over which it has some measure of control. Hence, it examines not only relationships between family members (O-O) relationships) or particular environments, such as the house and interiors (E-E relationships), but also the decisions and actions that occur as the family interacts with its many environments-natural and artificial"(Paolucci, Hall & Axinn, 1977:1).

This perspective highlights the changing nature of families and environments and offers a way to assess the dynamic nature of family ecosystems, and thus provide a base for policy, legislation, and program development to improve peoples' quality of life (Andrews, Bubolz & Paolucci, 1980).

Many of the ideas that lead to the development of next steps in family ecology emphasize Bertalanffy's (1968) general systems theory, information and communication theory, cybernetics (Ashby, 1958; Wiener, 1948;1967), Buckley's (1967) system's approach and Kuhn's (1975) conceptualization of a unified social science. In 1993, Michigan State University Professors, Margaret Bubolz and M. Suzanne Sontag (1993) made significant contributions in the development of the family ecology theory, which represents a "synthesis of assumptions, concepts, and propositions from ecology in several disciplines and from general systems theory, with its 'family roots' in home economics" (Bubolz & Sontag, 1993:424).

THE FAMILY AND GENERAL SYSTEMS THEORY

In family ecosystems, the family is seen as the fundamental human system made up of different subsystems. It is also understood that family members are joined in a network of pathways over which pass *materials*, *information*, and other forms of *potential energy*, in interaction with their environment. The family is composed of persons related by blood, marriage, or adoption, and also sets of interdependent but independent persons who share some common goals, resources, and commitment to each other over time (Paolucci, Hall & Axinn, 1977). Central in the model of family ecology by Bubolz and Sontag (1993) are concepts and constructs that integrate *family resource management*, *human development*, and *family relations*. Thus, the individual and the family are simultaneous focal units of analysis within contextual systems or environments.

The *environment* is all objects and forces external to the members of the family. This includes the *natural environment* like air, water, soil, temperature, gravity, topography, other life forms, as well as the *human built environment* and the *socio-cultural environment*.

Today there are many kinds of families: married couples without children, single parents with children, grandparents and grandchildren, couples without children, intergenerational families and so forth. In the United States there are very few families, if any, that completely correspond to the ideology and system of ideas, of the traditional nuclear family.

Historically, getting married and having a family served political, social, and economic functions (Coontz, 2000). The individual's needs and desires were not as important as serving the needs of the state and the family. Today, in the 21st century, there are new marital norms and values that make marriages more fair and, at the same time, also make marriage more easily dissolved. According to research findings, peer marriages are emerging where men's and women's roles are becoming increasingly more similar. The American family is an example of a *morphogenic system*, characterized by adaptive responses to changes.

STRUCTURAL COMPONENTS OF THE FAMILY

The structure of a family ecosystem is the pattern of organization described in terms of its systems, subsystems and their roles. What is called the family structural component is the manifestation of forces and mechanisms that interact and thus influence family processes.

A family can be made up of single parents with children, or two parent families with children, or a childless couple, or an intergenerational family, couples who cohabit, and so forth. Because of the various roles that family members engage in, either because of expectations or to accomplish a task, families all have unique patterns of interactions of matter, energy and information flows.

The entire web of family-environment relationships is intrinsically dynamic. In family ecosystems the reverse is true. What ecologists call parts are a recognizable pattern in an inseparable web. Therefore, family structures are manifestations of distinguishable patterns arising from the diverse characteristics of **individuals** and **families.** These patterns emerge due to their unique **structure of subsystems, ethnic and cultural origins, the individual's as well as the family's developmental life stage,** and **the socioeconomic status of the family in a society.** The individual and family attributes such as *needs, values, goals, resources* and *artifacts* also make up the dynamically intricate web of structural relationships.

Complexity and Hierarchy of Family Systems

There are numerous subsystems that make up the internal and external structures of the family. *Complexity* in families is shaped and determined by the number of individuals, which necessitates more coordinated functioning of a complex network of transfers of materials, energy, information, money, and processes. *Hierarchy* in families is the relationship between systems in which one is a subsystem or suprasystem relative to another system. For example, a sibling system is considered a subsystem in relationship to the family system. Because family systems are open, as well as information processing systems, they are adaptive and organizationally complex (Buckley, 1967).

Internal Structure of the Family

The *internal system* is described as regulating the relations within the family *transactions* between the family and smaller subgroups, such as husband and wife, sibling cliques, and the individual personality systems of family members. This organizes the internal structure of the family system, which is composed of subsystems of individuals and their roles and relationships with each other.

A *role* is defined as a set of system states and actions of a subsystem. A role occupant is the person that effectuates the behavior specified in the role. Roles themselves are pattern systems and hence do not interact. For example, an adult female in a family can be a wife, mother, daughter, older sister, younger sister, aunt, grandmother and/or great grandmother. Individuals have multiple roles. As a family unit *subsystem*, the *interpersonal subsystem* and the *personal subsystem* meet at the interface. Changes take place, which result in shaping and reshaping social space, physical space, and the thickness of the family system's boundary.

The family is organized with patterns of positions, roles, and norms. Individuals belong to different family subsystems in which they learn different levels of power and differentiated roles. For example, a married couple with no children may only have a *spouse subsystem*, consisting of marital partners, a husband and a wife. But a two-parent household with one child has an additional three parental subsystems. These include the mother and child, father and child, and the combination of parents and child. As the number of individuals in the family increases, the family system becomes increasingly complex.

External Structural Influences on the Family

There are also external family relationships that influence the structure of families. These are external systems for dealing with transactions between family and nonfamily events, or *noncollateral system*s like the school, occupational world, and marketplace. Because the family is an *open system*, it is sensitive to the external influences and transactions in order to maintain its family boundaries.

As family members are also acting and reacting parts of society, their social conduct can be analyzed as social roles following expected social norms and social patterns. These influences are, to a large extent, the family's social status and place in society. For example, a father who is a minister of a church, ministers to his congregation and thus has a prescribed position in the community. His role as a minister is a patterned system that has an influence on the family. Mother's external roles may be many, and the child's external activities may also be many, depending upon participation in the community. Therefore, the family's social status includes behaviors of social roles, social norms, and social patterns. Numerous subsystems make up the internal and external structures of the family, influencing the web of patterns that give structure to families.

Developmental Stages

The family is also a *bio-social system*. Certain physical and social tasks and accomplishments are important at various **developmental stages of individuals and families**. The developmental stages provide a road map or structure for the physical and social realization of the family's organizational pattern in space and time. Over time, individuals in a family undergo a process of interrelated changes as they interact and experience their environments. As a young child undergoes physical and social development, the nearby environment of the family is a most critical source in providing human and nonhuman resources.

The family constitutes a level of structural complexity that is called a *microsystem* (Bronfenbrenner, 1979). As the child progressively interacts outside the family with other systems, such as the school or peer groups, the interface between these two microsystems, e.g., the family and the school or peer group is labeled the *mesosystem* (Bronfenbrenner, 1979). Usually each member of the family has a different set of microsystems that interface called the mesosystem. Figure 3.1 shows the mesosystem for a child. For family members who are not directly involved in these contextual environments created by the interfaces of particular microsystems, their relationship to this context is called the *exosystem* (Bronfenbrenner, 1979). For example, the interface of the family and a mother's work place is her mesosystem, but for the child who is not directly involved in the interfaces of these two microsystem, this context forms the child's exosystem. There are also critical influences and reciprocal relationships with the institutions within society in the form of cultural expectations and values. These *macrosystem* influences are important to the development of the individual as well as to the family.

GOVERNING COMPONENTS OF THE FAMILY

Needs

Individuals and families have requirements that must be met at some level if they are to survive and engage in adaptive behavior. There are needs for having matter, money, energy, and information. There are needs for relating, being loved, being accepted, and communicating. There are needs for being, for growth and development, for self-fulfillment, for a sense of satisfaction, for the ability to control one's life, and for actualizing one's potential. In Maslow's (1968; 1975) system, human needs form a hierarchy in the shape of a triangle. Basic physiological deficiency needs form the foundation to the self-actualization needs at the apex. All humans have physical and security needs such as basic biological and physical needs for air, water, sunshine, food, clothing, shelter, and safety, but also important are psychosocial needs for belonging and love needs: for family, friends, affection, intimacy, respect, and esteem. At the apex of Maslow's system is self-actualization/spiritual needs, the need to understand the mysteries of human existence, the natural environment and our place in it (Maslow, 1954).

Values

According to Rokeach (1973), values are an enduring belief that a specific mode of conduct, of existence is personally or socially preferable to an opposite or converse mode of conduct or end state of existence. These values guide our everyday speech and conduct. Values make up what we are and propel us to action, or a particular kind of behavior and life. Humans choose values based on authority, on logic, on sense experience, emotions, intuition and the "science of the times." When someone tells you to "trust your feelings" this means that one's emotions are more valuable than authority or logic.

However, values, change over time and have a hierarchical nature, reflecting unique experiences and what is important at a certain time. For example, the concept of *familism* is extremely important to the Hispanic person whose value of the importance of family relationship often supersedes the value of an independent and autonomous individual. However, there are Hispanic individuals who have taken on the values of the importance of the independent individual from the context of the *macrosystem*. These outside influences, create within Hispanic families a network of relational strain and a need for reorganization and restructuring of previously held behavioral values of individuals and family relationships. To achieve a dynamic balance, family's adapt and change in order to achieve a quality of life.

Values can act as filters of information, determining to a large extent what flows in and out of the family boundary. Therefore, the family boundary is open or closed to certain influences from the outside as well as what is considered private or public family matters. Certain ethnic groups socialize their members to be similar to what marriage and family therapists label an *enmeshed family system*. In these ethnic families there are few outside influences. The family is described as a structurally rigid and a mostly closed system. In such family systems, rules govern, prescribe, and limit the individual's behaviors over a wide variety of areas. Other groups may socialize their members toward a *random family system*, which is loosely structured. In this structure, family members are often scattered by the wind and are very unpredictable. An *open type of family system* uses control strategies of equality, cooperation, participation, and consensus building. Meaning is derived from the practice of tolerance, dialogue, and pragmatism with authentic and expressive affect. Each type works, and no type is viewed as unhealthy or healthy (Kantor and Lehr, 1975).

Ethnic and Cultural Background

The **ethnic and cultural background** of a family is very important in influencing the *value orientation* of the family as well as the experiences encountered during a life time. There is ample evidence to suggest that human experience is a product of life conditions. Thus the immediate environment and the behavior patterns learned and practiced in one's ethnic and cultural family of origin influence who one is in the present context. A family's value system fits a particular culture and, as a functioning family, its interactions and behaviors fit together. They reflect the family values of class differences; religious differences; ethnic differences; whether one was influenced by a rural, urban or suburban life style; differences in the life cycle phases and, therefore, age of family members and their personal idiosyncrasies. The values and practices in which an individual engages are often a product of the experiences he or she has had in the family of origin.

The culture of families in poverty is defined by the lack of resources for a healthy physical development, and it places individuals at risk for unhealthy values and behaviors. Numerous studies (Oyemade, 1988; Martinez, 1988; Schorr, 1989) have emphasized the potentially devastating effects of poverty on human development. Schorr (1989) states that "Poverty is the greatest risk factor of all."(p.xxii). The culture of poverty is a greater and more decisive threat to families and society than differences found between ethnic or racial groups.

Goals

According to Kantor and Lehr (1975) there are three *resource channels*, *time*, *space*, and *energy*, that assist individuals in families in obtaining three desired universal human goals. These are *power*, which allows the individual to decide what he or she wants and the ability to get it whether it is money, goods, or skills. The second is *affect* which allows for intimacy and nurturance, a sense of loving and being loved. The last goal is *meaning*, which provides a philosophical framework that provides individuals with expectations of reality and helps to define one's identity so that we can begin to understand who and what we are. According to Bubolz and Sontag (1993), values and goals are the major motivating force in families and at times may exceed the resources within the families or may vary in time orientation.

DYNAMIC COMPONENTS OF THE FAMILY

Energy and Matter

All phases of any human phenomenon, including family life are accompanied by matter-energy and information transformations. There is *energy* in sunlight, sound waves, chemical reactions, magnetic fields, foods eaten, information heard, and books. There is energy in human services and in resources received, spent, or stored for future use, like money in a savings account. For example, in most kinds of household work or family relationship, one potential type of energy is transformed into another. There are many forms of *potential energy* involved in various processes within families. Potential energy is capable of driving processes in families. Energy transformation from one form to another results in what social scientists describe as individual and family processes. When energy is used, it no longer has potential for further work.

What is termed *high quality energy* takes many forms. Some are concentrations of actual energy such as high temperatures of the furnace or the cold of the refrigerator. Others are information containing objects such as the genes, computer programs, and political symbols. They all have in common the large amount of energy used in their generation and the potential for large amplification effects. Often an increased scarcity is identified with increased value. For example, ice in one's drink in the tropics is a high quality item, with high embodied energy flows. Why? Because refrigerators are required for making ice. The idea of energy-quality helps to explain why family systems invest in physical and psychosocial assets and store material goods to survive. One could say that the estimation of energy quality relates to the amount of maximum power that is derived from the energy source. The sun is of high quality due to the sum of heat that can be disperse by the sun. Similarly, the potential of embodied energy can be calculated by the use of input-output of dollar flows or the measure and the quality of information.

Social-emotional energy is more abstract. As energy input flows through the family, it may, in turn, activate decision making processes as its members process the transformation of the energy. Thus energy can influence social mechanisms within the family, such as learned skills acquired by a member of the family. Another example is continual chaos, an inability to adapt to change, or the failure to incorporate resources can lead to imbalances. Thus *fueling*, for example rest and relaxation, which varies from individual to individual, can become a focal point of family disagreements. One could stipulate that the quality of social-emotional energy investments made by families can be assessed and measured as the developmental outcomes of their children.

Because the family is an open system, it maintains its structure or pattern of organization through continuous exchange of energy and matter with its environment. The organizing activity of family living is a self-organizing system of human cognition or mental activity. This means that all interactions of individual and family life with an environment are cognitive and mental interactions (Bateson, 1979).

Energy in a family ecosystem allows family members to adjust behaviors and actions in relation to each other and environment. Sufficient energy is needed for open type families to survive. In such families members cooperate and participate in decision making. There are tolerance and dialogue within the family context. Members are responsive to each other, expressive, and authentic in their relationships. High value is assigned to individual initiative and responsibility (Kantor & Lehr, 1975).

Too much energy at a given point in time can cause adjustment problems in the family and lead to ruinous outcomes. For example, an unexpected fortune of winning a lottery, or inheriting a large estate can cause turbulence in the family ecosystem. The analogy of a boiling kettle of water, where increased energy is flowing through the system, applies to *atomistic families*. As energy is added as heat, the water molecules develop motion. The water boils faster, and the molecules tend to dissipate into the surrounding air as *entropy* increases. What is considered a liquid state is no longer present. Similarly, families experiencing a rapid increase of energy have difficulty maintaining stability. Winning the lottery can create a more atomistic family type, or a random type family, more prone to family disruption. Families have fuzzy boundaries, and energy is continually fluctuating in and out, which creates a family that is loosely regulated. According to Kantor and Lehr (1975), these are fragmented families, each member having little if any connection with others, favoring opposing values and behaviors that change or block what could become established patterns.

However, if insufficient energy flows through the family system, this can create a *closed family type*, sometimes called an *enmeshed family*. The structure is predictable and rigid. The subsystems seem to be "stuck together" with traditions and routines within the family. This usually occurs in families with boundaries that are less permeable to permit adequate energy exchanges. In such families space is fixed and carefully guarded.

It is important to understand that all living systems are *open systems* that continuously exchange energy and matter with environments. However, some family systems' energy flows are routinely steady, emphasizing family privacy and boundaries (Kantor & Lehr, 1975). The family is about preserving tradition and continuing the past into the future.

Information

Information is a process relationship (sender-channel-receiver) that carries meaning and gives structure to matter-energy flow in the interests of a goal (Bubolz & Sontag, 1993). The addition of more information content or energy flow through a family system produces more structure with complexity and information (Odum 1994). Thus, information is an additive function of complexity. We see a rapid increase of information flow in the 21st century.

It is sometimes thought that accumulation of knowledge allows more to be done with fewer resources. Today there are huge amounts of stored information, subsequently large amounts of energy are required for replacement and maintenance of these stored information. The potential problem as we saw with the Y2K episode could have been disastrous. Families are confronted with a great variety of information, from experts, TV, neighbors, signs on billboards, radios, songs, and the Web. Too much conflicting information can be associated with uncertainty and, therefore, leads to the possibility of greater disorder.

Money

According to the ecologist, Odum, (1994) money is a type of energy flow, since it is a quantity that controls or releases other energy flows. In terms of quality, it is considered high value in its interactions, because of its magnifying power and its amplifying effect. Families use money as an exchange medium. Money flows countercurrent to materials, information, and other matters used. When an individual pays for food, there is a price attached to the purchase. The price, in turn, has a regulator mechanism operating. This may often be externally set by the costs of production and marketing.

Human and Nonhuman Resources

The basic forms of resources - matter, energy, information, and money - are converted to be useful to a family for attaining a goal (Paolucci, Hall & Axinn, 1977). There are human and nonhuman resources. Human resources include abilities, knowledge, skills, human energy, and time. Nonhuman resources include money, material objects like artifacts, and many human constructed and natural environments. Many of the resources can be exchanged, for example, goods and services can be exchanged for money. However, not all resources are interchangeable, nor are they renewable. There are resources that are scarce and limited. Oil is a vivid example of a scarce nonhuman resource that is needed to keep technology, as we know it, running. Human love is a particularistic resource that is not easily exchangeable with other human resources, except for love.

Families extract and transform environmental energy such as fossil fuel sources for energy, symbolic resources such as income and education, human energy in the form of skill and abilities, and material resources. All family behavior carries an energy price tag. For example, the food eaten by individuals in the family is transformed into caloric energy, which forms the basis for all human activity. Therefore, the structure and quality of family interactions are related to energy resources as well as the family's ability to extract such resources. One could postulate that a happy and stable family probably does better in securing resources. Income and family size affect energy consumption, and richer families tend to use more energy than poorer ones.

A fundamental responsibility of families with children is to help socialize children to develop their human capital (intelligence, skills, abilities, talents, a caring nature) or human resources. Human resources can then be transformed into social capital (volunteering, social supportive services, building a safer environment) for the good of society. Families can be best understood from the nature of the units and relationships as they manage resources to attain goals.

INFORMATION PROCESSING COMPONENTS OF THE FAMILY

"Adaptation is behavior that in some way changes the relation of a system to its environment, whether by altering itself, the environment, or both" (Kuhn, 1974:484). Therefore, what is called a successful adaptation is one that increases the likelihood of achieving some goal of the system.

Perception

Perception is the process we use to make sense of our experiences. This perception/action process picks up information through seeing, hearing, touching, and making sense of the sensory information in an active and exploratory sense. Individuals actively select or choose what they will focus on and then organize the stimuli in order to interpret the sensory data. Each family member acts on an interpretation of what the environment appears to be and not necessarily on the basis of objective reality of the physical or social environment. What is sometimes perceived as an opportunity by one member of the family may not be seen as such by another. Each member is an active participant or actor who is guided by unique perceptions and actions throughout personal development. Individuals choose an event to perceive. We usually choose stimuli that are more intense than others, or that reflect our motives or interests more than others. For example, we will pay attention to a sudden loud explosion, or overhear what mom might be saying about us to her friend on the telephone in another room.

According to (Gibson, 1997), an ecological psychologist, there are several factors that influence perception. The person's goals, interests, and abilities; the "seems to be like" environment encountered; the fit or match of the person's perceptual state to take advantage of the opportunities of the environment; and the individual's actions in response to such perceptions. This means that individuals respond selectively to their environment and then make interpretations through personal and cultural meanings (Boss, 1987).

Individuals in a family are constantly making judgments about others, deciding what they are like, or what they will do, and then providing explanations for their behaviors. Individuals make inferences and believe they know their family and friends well. Often the family's expectations, stereotypes, and beliefs of relationships influence the individual and vice versa. The social context as well as the cultural setting of the perceiver and the event influence the formation of the impression. Therefore, meanings assigned to experiences are influenced by a number of factors that affect our responses such as culture, roles, biases, emotional states, and past experiences, as well as physical and social limitations.

Communication in the Family

According to communication theory, all human behavior is communication. All communication, whether it is a dyadic communication, between two persons, or whether it is within a small group (face to face), such as within the family, or whether it is a form of mass communication, which is often mediated through print or electronic media, all share in common the process of creating meaning between two or more people. If there were a perfect sharing of meaning, the receiver's meaning would be the same as the sender's meaning. However, verbal messages are influenced by culture and connotations, as well as by private meanings. There is a need in families to achieve a shared meaning, so there is some correspondence between the message as perceived by the sender and the receiver.

In families there are different levels of communication skills. Young children, who cannot take the role of others, tend to speak to others uniformly, making few accommodations to different receivers. Older children are better able to adapt patterns of speech to a variety of listeners. Adults can effectively convey complex meanings, including references to people, things, and situations that are not present. A problem can arise when people use abstract language or double bind messages (a hidden message intended for the receiver), which frequently cause communication difficulties that have to do with vagueness in interpretations. Sometimes individuals will use euphemisms to substitute for emotionally charged terms rather than using blunt words. An elderly person might regard terms like living will or graveyard as blunt.

Ways in which families use time, space, and energy to create a network of interconnected relationships are referred to as *target dimensions*. In most family conversations are intended for the individual to achieve a target, such as *power*, *meaning*, and *affect* (Kantor & Lehr, 1975). In order to understand what the target may be, a useful question to ask is, "What are family goals?" Power can be described as the freedom to try to obtain what is wanted in terms of money, good, skills, and a sense of efficacy. Power ranges from freedom to restriction. Meanwhile if the goal is affect, then the goal may be a sense of intimacy or nurturance (joining) or separating. If the conversation builds on a relationship based on meanings derived from sharing or not sharing, it may reveal information about identity or philosophical beliefs.

INTERRELATIONSHIP COMPONENTS OF THE FAMILY

Decision Making in the Family

According to Bubolz and Sontag (1993:436), "Decision making is the central cybernetic control system of family organization", which directs actions for attaining individual and family goals. Where there are rules of negotiation, a family decision making process involves interaction, communication, and weighing of alternatives. This process reflects individual perceptions, needs, values, as well as family needs and values and reflects a greater family input and accommodation. Wise decision making in families is crucial to creating an environment that will aid individuals to reach their potential.

A stressful situation often involves conflict of values of self-interest versus family welfare. Due to limited resources and differing views, the decision making process may be a reflection of authority and power relationships within the family. *Power* is the potential ability to influence family members, and *authority* refers to beliefs concerning the proper exercise of power (Safilios-Rothschild, 1970). Who makes the decisions in families depends upon family expectations concerning power, who should decide, age, competence, sex, and other factors (Paolucci, Hall & Axinn, 1977). When decision making is based only on self-interest, there is a potential danger of a zero-sum power confrontation in families. If one person wins, another loses.

Decision making functions in the family stabilize and maintain the family's most important values as well as bring about nondisruptive change to adapt to changing environments. Because families have a place and time in history with certain functional expectations, some changes are mandatory and are imposed. In such cases, decision making can be based on the family's image of what it feels it should do and what, in reality, it can and does do to ensure survival and status.

Decision making rules in families consist of methods in which alternative courses of action are evaluated. There are usually three decision making rules in families. *Preference ranking* is subjective ranking by the decision maker based on his/her preference or criterion. Several dimensions are ranked from best to worst. *Objective elimination* is quickly recognized because it is based on limits imposed by the environment and is based on objective rather than subjective conditions. In such cases, there is no consistently best alternative. Therefore, there are fewer dimensions, and there is less complexity when making a decision. There is a possibility of making sacrifices and at the same time being satisfied with what is decided as important or necessary (Plous, 1993). *Immediate closure* occurs when only one action is focused upon. There is no ranking or elimination of alternatives, but a single course of action is immediately accepted. There is no debate, no compromise, no fact finding that leads to the action. This process is quick and often followed by a rationalization after the decision has been made (Paolucci, Hall & Axinn, 1977:95-96).

Factors Affecting Family Decisions

Factors that affect family decisions include the communication skills of individuals, the age and gender of family members, the alliances that are found over issues, the problem solving skills of individuals as reflected in their thinking styles, and the information and objective data gathered and presented for discussion. As families go through the stages of family development, various individual and family roles and responsibilities also influence family decisions.

Three Kinds of Thinking Process

Within families, there are three kinds of thinking processes that could influence the ways decisions are made. There may be those whose process of thought reflects *mechanistic thinking, intuitive thinking*, or *systems thinking*. Each type arrives at a solution very differently. Mechanistic thinkers tend to follow predetermined, orderly, step-by-step approaches to reach a solution. Intuitive thinkers rely on hunches and perceived patterns rather than strictly rational approaches. Individuals who use systems thinking examine the whole in all its complexity rather than parts of the whole. From this point of view, a part can only be understood by seeing the whole (Senge, 1990).

Individual members of the family may arrive at solutions differently. In order to achieve a desired family goal, individual needs and strengths of others in the group must match the strategy used or the decision made. How families make decisions, who is involved in the decision making process, and what is decided upon may depend, in part, on the conception the family has regarding desirable outcomes. What a family does is usually what it believes will ensure the family's survival and place in society. The decisions made in families can shape individual and societal value systems, and can influence the use of human and nonhuman resources in improving the quality of life, as well as in preserving the natural environment.

Sustenance Activities

Families engage in sustenance activities in order to meet needs and to work toward survival (Bubolz & Sontag, 1993). All activities for production, consumption, and nurturance that take place in the household require the transformation of matter-energy, information, and money. Families engage in activities that support the necessities of life such as food, shelter, clothing, and safety as well as social-emotional needs. Most of these activities are repetitive and cyclical for a time period. Others may be more enduring over the developmental life course of the family. For example in certain households, the morning starts with everyone getting up for work, school, or for doing household chores like making breakfast for everyone. This is a repetitive cycle that occurs every morning for five days a week with Saturday and Sunday usually presenting different patterns of household activities. The economic livelihood of a family is sustained by going to work. Engaging in this pattern of activity brings in added potential energy for exchange, money. Money can be exchanged for almost any other resource except love. Therefore, by engaging in the economic livelihood of a family, which is one form of sustenance activities, the basic needs of the family are met, and values are realized. The family's quality of life or their well being is enhanced.

LIFE PROCESS

Family Adaptation Involving Throughput Activities

Adaptation is behavior that in some way changes the relation of a system to its environment, whether by altering itself, the environment, or both (Kuhn, 1974)

Therefore, what is called a successful adaptation is one that increases the likelihood of achieving some goal of the system.

The family is a complex organization. By organization, we mean the structured relationship between parts of the whole. A fundamental assumption is that we are speaking of a whole composed of parts. A complex organization is the consciously coordinated actions of two or more parties toward the joint effectuation of a goal (Kuhn, 1974). It represents the adaptive means in which interdependent organisms (in this case the individual/family) are able to meet a variety of goals in diverse environmental conditions as they change and adapt over time (Micklin, 1973).

In the family, matter, energy, and information are transformed through choices and managing processes by differentiated family tasks (associated with family roles) to work toward a specific goal of maintaining and sustaining family well being. Family roles and role behaviors are more or less normatively defined and expected by society. The normative dimensions of roles, role behaviors, and role patterning and functioning are joint activities toward achieving the goals of family systems.

THE NATURE OF TIME IN FAMILIES

As we examine time as a construct, we can see the structure of time as a succession or as duration. For example, an individual's orienting characteristic reveals what is important to that person. By selecting, directing, and maintaining attitudes and behaviors toward the past, present, future, or nontemporal realms of experience, a person is emphasizing one or more of these realms and the particular importance of the relationships among them. *Past orienting* is remembering or reenacting something that has already taken place. Past orienting is always concerned with history, especially family history. *Present orienting* is the importance of the here and now. What people are actually sensing, feeling, experiencing, and doing is considered most important. *Future orienting* emphasizes what is to come by anticipating, imagining, and/or planning for it. *Nontemporal orienting* involves events unrelated to calendar time, such as fantasizing, dreaming, meditating, when unbounded by past, present, or future calendar time constraints (McCubbin & Thompson, 1986).

If we look at time as a flow, we see that it can be *bidirectional time* or *linear time* or *unidirectional time*. For example certain family episodes may be recurrent and are, therefore, considered to be bidirectional. In the study of human development, time is usually seen as a unidirectional flow or linear phenomenon. A family genogram is a tool that depicts the relational properties of families over the generations.

Successful families integrate the submechanisms of time orienting. This enables families to organize their experience of past, present, future, and nontemporal events into a pattern for interrelating events experienced in different time spheres. For example, *family traditions* are those events, activities, and practices that families have done in the past, and which they are likely to continue to do because of their value or respect (McCubbin & Thompson, 1987).

Family *patterns of stability* involving traditions, celebrations, and family routines are an essential part of family life. As these events occur there are feedback loops into the family system, which allow for a bridging of generations to establish continuity. Though culture is always changing, in the midst of transitions and change, traditions and celebrations help families to negotiate the transformations that are taking place. According to studies by McCubbin, McCubbin, and Thompson (1987), there is a relationship between family satisfaction, family celebrations, family bonding, and family coherence.

DIMENSIONS OF SPACE

Our psychological well being and physical well being are connected to happy, beautiful, and healthful places. Our home is such a place, where we pay attention to the air, the temperature, and the artifacts that express our identity. We come to be "addicted" to our homes, because they support our families' physical and social bonds for each other. Homes also buffer us from the outside influences and commotion that we do not wish to invade our privacy. Homes also contain objects that help us to find meaning and to express ourselves (Gallagher, 1994). For families the home represents a concrete physical space as well as an expression of psychological space. How a geographical area or space will be used in a home is regulated for social life. For example, there is Dad's chair or children's playroom, as well as locked doors that convey markers for privacy or personal space and territory.

Individuals and families structure and regulate their interactions spatially. According to Lyman and Scott (1975), there are four types of *human territories*. *Public territories* allow for freedom of access for anyone. *Home territories* are public areas taken over by groups or individuals like gang turfs (territory). Social gatherings occur in *interactional territories* with marked boundaries and rules of access. *Body territories* encompass the body such as personal space, body images, and psychological self. According to Stea (1970), human territorial spaces should be thought of as discontinuous. There are territorial spaces for work, one for recreation, and one for family life, around which social life is organized.

A variety of *offensive moves* and *defensive moves* take place to keep a territory like personal space secure. Factors that influence personal space include one's culture, level of acquaintance or intimacy, gender, status, topic or task, feelings, and personality. It is also understood that family members at different times have shifting needs for connectiveness and separateness as reflected in emotional and physical distance regulation.

The personal space bubble of the family is the home. Today the home is spatially designed to meet the demands of the activities that occur in that space. For example, a kitchen is for eating, the bedroom is for sleeping. In some families, privacy is not valued. What is meant by privacy is the selective control over access to the self or to one's group. In the United States there the preferred home is a single family dwelling. This type of dwelling represents a form of managing the boundaries of that space.

The amount of space a family occupies reflects, to a greater extent, the amount of resources the family has. A family's easy access to resources is often related to access of power in society to influence the direction of energy flow and interactions. As families interact with their environments, they are entrusted to monitor and preserve a human ecosystem that will maintain a quality of life for all.

THE POWER OF THE THEORY: FAMILY AS AN ECOSYSTEM

Ellen Swallow Richards, Beatrice Paolucci, Margaret Bubolz, M. Suzanne Sontag, and many other women scientists have contributed to the development of human ecology. They have identified the family ecosystem as the fundamental and basic ecosystem of human ecology. In the family environment, individuals are physically cared for and socially nurtured to become the organism that interacts with the environment to create a web of interconnections called human ecology. This is a science built on values that will enable individuals to actualize to their fullest potential and contribute to a quality of life, while at the same time protecting the environment so it can sustain life. Life begins in the family, which can be a beautiful, healthful, and socially nurturing place in interaction with the natural environment, socio-cultural environments, and human built environments. The family is the first as well as the fundamental human ecosystem.

As human ecologists we are told to think globally and act locally. Acting locally can start with the most basic human ecosystem: the family's interaction with the environments. All components of a system as identified earlier in this text are reflected in the family. The family system as a whole is greater than the sum of its parts. The power of the future is in the hands of the family. As families make decisions and manage their resources, guided by individual needs and values, they influence the network that makes a high quality of life possible.

Figure 3.1

Mesosystem For Child

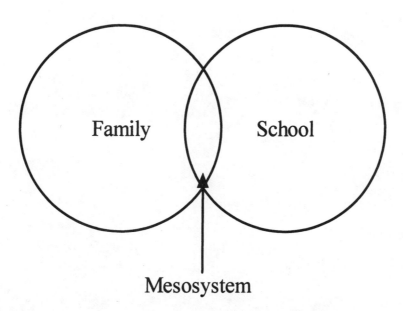

Exercises
Chapter 3 CONCEPTS

This chapter on family ecology presents many essential concepts. While all are important, perhaps you see some as more important than others. List and define what you see as the 5 most important concepts in this chapter.

1.

2.

3.

4.

5.

Chapter 3 EXAMINING THE CONCEPTS

Choose the three most important concepts or ideas in this chapter on family ecology. Why you believe they are the three most important concepts or ideas concerning family ecology? As you prepare your rationale, compare and contrast terms and ideas.

Chapter 3 APPLYING THE CONCEPTS

Two important concepts in this chapter on family ecology are communication and values. Choose two other important concepts from this chapter. For each of these two concepts, write an example to show how it can be applied. Give the page number where your application would fit in the chapter.

1.

2.

Chapter 3 SYNTHESIS

This chapter focuses on family ecology. Family ecology is described in terms of many concepts and ideas. Now that you have read the chapter, write this chapter's final paragraph. In this paragraph, draw conclusions and link together what you regard as the most salient thoughts and concepts.

Chapter Four

Psychological Ecology

At the beginning of the 21st century, psychology has become strongly concerned with the environments of behavior, including internal environments such as the nutritional environment, the neurochemical environment, the structural environment of organs, muscles, bones, and other structures, and, of course, genes. These insights have clarified how behaviors are affected by neurotransmitters, such as dopamine, histamine, and serotonin, testosterone, estrogen, and progesterone (Pert, 1997).

This emphasis on internal environments may be seen as a relatively new perspective on what has been, for psychology, a dominant focus on the internal dimensions of human reality. In his popular review of the importance of psychological environments, Gross (1978) sought to show the influence of psychology on contemporary culture. This influence has become even more evident in recent years.

PSYCHOLOGICAL ENVIRONMENTS

In various ways, the focal point of much of psychology has become the psychological environment. The psychological environment - what is said to be inside the person - has been described in theories in a multitude of unique ways.

Freud and Psychological Environment

In the 20th century, psychoanalysis led the way toward defining the concept of psychological environment. Freud (1953; 1960, 1961; 1965) dealt with homeostasis as a central assumption of his conception of human nature. For Freud, the psychological environment had a topography consisting of the *conscious* mind, the *preconscious*, and the *unconscious*. The psychological environment of which we are aware is, in these terms, that which is conscious. That which we can become aware of by remembering is the preconscious. The psychological environment of which we remain generally unaware is the unconscious mind. Here is material that cannot readily become part of the environment of which we are conscious. The unconscious area contains psychic energy, and is the heart of drives.

Freud made an important distinction between psychological environments of which we are aware and psychological environments of which we are not aware. Thus in psychoanalytic theory, the psychological environment of the developing individual does not necessarily incorporate memory of early developmental experiences. It includes the processes of behavior and interaction in which the person is presently involved, and which form the basis on which the person develops ego strength and self-awareness. It includes self-awareness in the present moment relative to drive management, self-management, the management of reality, and the internalization of rules and moral standards.

Freud's description of this conceptualization of an internal psychological environment was not, in all respects, completely new. However, he was able to formulate it and popularize it. Now, a century later, his notions of psychological environment and topography of mind remain a part of popular culture.

Erikson and Psychological Environment

In the psychosocial approach of Erikson (1959), the psychological environment is experienced in the form of conflicts that are sequentially situated in zones of the body. Zones of the body are experienced differently in the process of psychosocial development in a series of stages

Jung and Psychological Environment

Jung (1954) described personality structure quite clearly recognizing a psychological environment. For Jung, the ego is the conscious part of the personality. His view of the unconscious was quite different from Freud's view. Jung saw the unconscious as incorporating a potent force known as the *collective unconscious*, which is full of ancestral memories in the form of archetypes. Archetypes are also called *images* or *imagoes*. They are forms of universal experience that can constantly influence one's actions. Examples are the *mother archetype*, *hero archetype*, *shadow archetype* (residue of animal nature), the *anima* (feminine archetype) and *animus* (masculine archetype).

Bandura and Psychological Environment

Psychological environments are prominent in the cognitive learning theory of Bandura (1977, 1978, 1986). His model of triadic determinism, involving *environment* (stimuli), *behavior* (responses), and *person* (biology, cognition, motivation, perceptions, and needs) incorporates ways in which persons differently perceive and construct environments. He emphasizes modeling and the cognitive representation of experience, including representations of others and consequences of their responses. His view is that cognitive representations influence learning and behavior. Cognitive contents are encoded symbolically and organized. These contents, which may be thought of as resident in the psychological environment of the person, can then be used in processes of behavioral rehearsal, and in the production of behavior.

Behavioral production depends on motivational processes as well, and this relates to how the person represents or constructs the consequences of his or her own actions, as well as the observed actions of others. These constructions of behavioral consequences can influence whether an act is inhibited or disinhibited. Persons construct symbolic psychological environments involving representations of gender, representations of aggressive behavior, representations of cooperative behavior, representations of self-control, representations of forethought, and representations of self-efficacy.

How one constructs psychological environments has much to do with actions, and those actions can include choosing future environments from which psychological constructions are more or less likely to occur. For example, students choose environments that can facilitate or hinder high academic achievement. Choosing an environment in which symbols of academic mastery are more prominent thus symbolized in the psychological environment can foster academic and career success. More generally, persons can symbolize experience in ways that lead to feelings of mastery in virtually all aspects of life. These *self-efficacy* constructions can be very influential relative to many aspects of self-control (Bandura, 1982, 1988, 1992, 1993, 1995).

Humanistic Views of Psychological Environment

In existential and humanistic theories one finds a rich conceptualization of psychological environments. One experiences the self in the world, responsible for constructing an unfinished self and for creating the meanings of one's experience. For Frankl (1962; 1997), the process of making decisions is central to the making of meaning. Maslow (1968; 1975) described a process of *self-actualization*, in which it would seem that self is particularly in touch with the psychological environment. Rogers (1961; 1980) also framed self knowledge in terms of self-actualization and openness to experience. Processes having to do with the self are valid to the extent that one posits some sort of psychological environment as a representation of reality and personal meaning.

New Age and Psychological Environment

The concept of psychological environment as consciousness also is prominent in "new age" consciousness. As Ferguson (1980) explained in *The Aquarian Conspiracy*, there is a new consciousness of transcendence, in which new meanings of human development forged. More recently Deepak Chopra (1989; 1993) has discussed a consciousness unlike that normally associated with a Newtonian objective world. He conceptualizes a consciousness in which mind and body are seamlessly integrated.

Psychological Environment from Piaget's Perspective

Piaget's theory of cognitive development emphasizes psychological environments in the form of constructions of knowledge (Piaget, 1950, 1963, 1972, 1973). For Piaget, knowledge was not simply acquired as progressively more complex collections of facts. Knowledge was constructed as a product of mental actions in the course of processes of *adaptation* and *organization*. Thus, int his view, children are not smaller adults who happen to have fewer facts. The reality of children is based on constructions of reality that are qualitatively different from the constructions of adults. Their cognitive structures are different from those of adults. As new concepts emerge, they emerge from construction, not from acquisition. A unit of cognitive representation in the psychological environment of the developing, self-constructing person is known as the *schema*.

Vygotsky and Psychological Environment

Vygotsky emphasized the importance of social and cultural environments on human development (1962, 1963). According to Vygotsky, action influences thinking. Action is shaped by cultural symbols, and competence is defined in terms of mastery of what is prescribed within a culture. For Vygotsky the focus is on representations of the effects of culture on the developing individual. Internal processes of signalization involve the use of tools to control external environments and signs to control the self. The psychological environment incorporates these forms of symbolic content, which are directly associated with one's developmental experiences.

ORIGINS OF BRONFENBRENNER'S MODEL

Psychological Environments

The work of Lewin (1935) was a major influence on psychological ecology, and in particular Bronfenbrenner's model. Lewin's project was discovery of a natural science and natural laws of person-environment relationships with a particular focus on the individual's life space or psychological environment, but with attention as well to the relationship of the individual to natural and social environments.

Lewin conceptualized an internal environment consisting of *life space*. He saw development as involving the expansion of life space in terms of greater scope and depth. The development of the person involves the development of life space, as certain areas become more accessible and others less influential. Life space incorporates dynamics of motivation. Thus if one understands a person's life space, one can describe and explain a person's behavior.

Nonpsychological Environments

The effects of the nonpsychological environments on behavior and development are evident in environmental psychology, which is a basic foundation of Bronfenbrenner's work. Environmental psychology is said to incorporate the study of relationships of human behaviors to the environments in which people live (De Young, 1999), often called *behavior settings*. Particular behaviors become associated with settings, for example libraries, college classrooms, grocery stores, or automobiles. Some settings afford behavioral flexibility and choices. The classic work of Barker (1968) and Barker and Wright (1951) are illustrative of this focus in environmental psychology.

Bronfenbrenner's (1979, 1989, 1999) form of psychological ecology evolved from this work. His work also was a reaction to the insufficiency of the experimental child psychology of the 1960s and 1970s. In the 1970s there was an emergence of the view that in the future it would not be acceptable to conceptualize the development of the individual apart from the contexts in which development occurs.

Bronfenbrenner's model reflects Lewin's (1935) view that behavior is a function of the person and the environment. The environment in this case refers to the real environment in which the person lives, rather than contrived and presumably controlled laboratory environments.

BRONFENBRENNER'S ECOLOGICAL SYSTEMS

The essence of Bronfenbrenner's (1979, 1989) approach to describing an ecological systems theory consists of his levels of ecological analysis. These are a series of systems that are regarded as useful in understanding behavior and development.

Microsystem

Bronfenbrenner (1989:227) defines *microsystem* as a "pattern of activities, roles, and interpersonal relations." They are experienced in face-to-face settings over the course of an entire life. These microsystems contain physical and material features as well as other persons. However, microsystems are not simply physical places or other people. Bronfenbrenner views the family as an important microsystem. The family is not simply defined by the physical setting. The essence of the family as a microsystem is the interaction that occurs there. School classrooms and peer groups also incorporate microsystem relationships. Participants in these interactions possess the characteristics that are the outcomes of their developmental histories in these systems.

Mesosystem

"The *mesosystem*, comprises the linkages and processes taking place between two or more settings containing the developing person" (Bronfenbrenner, 1989:227). A frequently cited example of a mesosystem is the link between the home and the school. The child participates in a family microsystem and a school microsystem at the same.

Exosystem

An *exosystem* is defined by the linkages among two or more settings. The exosystem does not contain the developing person. However, events that occur in an exosystem influence processes that occur in a person's immediate setting (Bronfenbrenner, 1989:227). For a child the parent's workplace could be regarded as an exosystem. A university Board of Trustees can be regarded as an exosystem for college students.

Macrosystem

Bronfenbrenner's definition of *macrosystem* reflects the importance of patterns of micro-, meso-, and exosystems characteristic in a culture. He emphasizes the impact of culture, subcultures, and broader social context, as they relate to "developmentally-instigative belief systems, resources, hazards, lifestyles, opportunity structures, life course options, and patterns of social interchange that are embedded in each of these systems" (Bronfenbrenner, 1989:228). One might think of macrosystems as software that runs other systems. Macrosystems are always changing. Each generation experiences different macrosystems.

Chronosystem

Bronfenbrenner (1989) discussed a *chronosystem* model, in which attention is focused on changes in context are associated with time. With the passage of time there occur important events that bring about changes between the person and environments. Bronfenbrenner and Morris (1997) note that historical time is a major influence on shaping the life course. The order of experience, succession and duration of events, and how events are associated over time are all influential.

The notion that *cohort* or *generational influences* have powerful shaping influences on the life course is known as *macrotime* (Bronfenbrenner & Morris, 1997).

The powerful effects of cohort or generation have been recognized by other human development theorists (Erikson, 1959). These effects are found as basic developmental experiences in lifespan human development texts. There is indeed something special about the historical moment in which one's life occurs. There will there never be another cohort that experiences the events associated with, for example, the "Boomers" or the "Generation Xers".

Bronfenbrenner's emphasis on time is not limited to the effects of cohorts. Bronfenbrenner and Morris (1997) discuss the effects of *microtime*, which refers to patterns of continuity and discontinuity in the life course, and *mesotime*, which refers to periodic changes.

ECOLOGY AND DEVELOPMENTAL OUTCOMES

How do ecological systems influence developmental outcomes?

Influences of Microsystems

Proceeding through Bronfenbrenner's ecological systems, one could ask the following questions about microsystems.

- Does the developing person experience adequate support and the necessary range of experiences at appropriate times?

- Are the interactions experienced in microsystems optimal for positive development?

- Are there adequate experiences with symbols and objects?

In Bronfenbrenner's view, proximal processes are essential to the process of development. Without adequate proximal processes, optimal developmental outcomes will not occur.

Influences of Mesosystems

Moving on to mesosystems, one might ask the following questions.

- How effectively are the developing person's microsystems connected?

- How frequently are the person's microsystems connected?

- Do these connections serve positive developmental functions for the person?

- Are the microsystems congruent in their emotional characteristics and in their prevailing values?

- Do contributing microsystems represent conflicting expectations for the developing person?

- Is it easy or difficult for the developing person to make transitions across various microsystems?

Influences of Exosystems

Certain questions can also be asked about exosystems.

- Are exosystem institutions adequate and developmentally supportive?

- Whom do they support?

- Do institutions support positive development of children, adolescents, and adults?

- Are institutions adequately staffed?

- Do exosystem institutions actually produce developmental outcomes that are directly contrary to the program mission and purpose?

Influences of Macrosystems

Turning to the developing person's macrosystems, other questions emerge.

- Do prevailing cultural elements and patterns foster positive or negative developmental outcomes?

- Is there an ambience of violence?

- Is competition a basic value in the person's macrosystem?

- Do the prevailing values reflect racism, sexism, or ageism?

An Example of Influences of Ecological Systems

We will consider the ecology of child maltreatment as an example of how ecological systems can be used to examine an individual's developmental experience.

- In the family microsystem, are parental expectations for children realistic or unrealistic?

- Do parents' expectations match a child's abilities?

- What are the parenting actions of parents? Do they rely on punitive control techniques?

In examining the mesosystem as it might be related to child maltreatment, other questions emerge.

- To what extent is the family connected to other systems in the community?

- Is the family isolated?

- Is it possible that child maltreatment might go unobserved due to isolation of the family?

- Do patterns of family mobility prevent a family from becoming a part of the community?

Questions relating to exosystems are prominent in the study of child maltreatment. The exosystem incorporates local, state, and federal bureaucracies that are in place with a mission to deal with child maltreatment.

- Are social organizations and institutions really effective?

- Are their interventions appropriate?

- Are these organizations staffed by well-trained professionals?

- Are family professionals prepared to be effective in dealing with child maltreatment?

- Is the legal system effective in dealing with cases of alleged child maltreatment?

- Is the judicial system effective in cases of child maltreatment?

One could add that the macrosystem also has much to do with how a family experiences culture as well as what is experienced. A prevailing culture of violence might shape a higher incidence of child maltreatment.

ECOLOGICAL SYSTEMS AND DEVELOPMENT

In ecological perspectives, one does not necessarily find a strong emphasis on the orientation that is known as *developmentally appropriate*. Rather than orderly and predictable sequences of quantitative and qualitative developmental change, one finds a general concept that the characteristics of environments and persons are mutually interactive. Thus, the progression of development depends substantially on these patterns of interaction.

Some elements of contemporary ecological perspectives appear to be comfortable with a behavioristic emphasis on the impact of environments on behavior, while diminishing attention to the reciprocal interaction of persons and environments. This one-way street concept is a limited view of the ecology of development and, if taken to extreme, can cause one to focus on how certain environments create victims. While it is certainly true that some environments, natural, human constructed, and social, can be pernicious as developmental influences, this is only one facet of the mutually influential process of person-environment interaction.

Whether a child lives in a safe neighborhood or a dangerous neighborhood, or associates with poor people or rich people, can certainly shape developmental outcomes. However one cannot discount the potentials of individuals to choose environments and to shape them as well. Thus, a human ecological point of view is not a human science of victims of circumstances.

Bronfenbrenner's model emphasizes the collaboration of the person and environments. He views the developing person as constantly growing and active. Persons and environments are always changing, and there is a mutual accommodation of persons and environments, which continues over the course of the entire life span.

THE INDIVIDUAL IN BRONFENBRENNER'S MODEL

Individual differences are important in Bronfenbrenner's model as personal characteristics that may facilitate development or serve as obstacles in interaction with environments (Bronfenbrenner, 1993; Bronfenbrenner & Morris, 1997). Personal characteristics can do this by shaping and influencing proximal processes. Some characteristics may draw out and amplify the effects of certain environmental characteristics. Persons who are active, strongly assertive, very attractive or unattractive all tend to elicit certain types of responses from social environments. Persons who tend to be highly controlling may tend to seek environments that can be controlled or environments that are structured consistent with their preferences.

A BIOECOLOGICAL MODEL

In recent developments of a bioecological model, one sees in Bronfenbrenner's thinking an enhanced role for personal characteristics. This seems consistent with Lewin's views concerning human development as a biological science of organisms with environments, focusing on phenomenology, motivation, and action of the person in life space.

In view of Bronfenbrenner's evolving view, it would seem useful to focus on how to positively expand the developing person's life space. Thus it would seem useful to focus on enrichment of experience and developing a larger and more diverse, more complex, and more differentiated life space. As the individual continues the developmental process of constant change, there are microsystem opportunities relative to new and more developmentally appropriate processes that can be accessed by the active, changing person. Mutual and progressive accommodation can occur most effectively when the developing person can select from progressively evolving environments to achieve a goodness of fit.

APPLICATION: RISK AND ASSET APPROACHES

Bronfenbrenner's model is regarded as one of the foundations of epidemiological risk models, in which anomalies of personal and social developmental outcomes are analogous to heart disease, diabetes, or other maladies (Bogenschneider, 1996). In these models, developmental outcomes flow from a balance of risks and other factors, for which statistical vulnerabilities can be established in definable populations.

A model developed at The Search Institute incorporates assets such as family support, positive family communication, caring neighborhood, safety, a community that values youth, achievement motivation, and integrity (Scales, 1999). Based on these data, it is concluded that having more risks is associated with more problem behaviors, and that vulnerable youth can benefit from assets.

In recent years there has been an emphasis on asset models and somewhat less emphasis on developmental risks. The role of developmental assets has been studied, for example, with regard to adolescent drug use (Block, Block, & Keyes, 1988; Brook & Newcomb, 1995; Brook, Brook, Whiteman, Gordon, & Cohen, 1990; Brook, Whiteman, Gordon, & Cohen, 1986). Asset approaches to human development (Roehlkepartain, 1997) are based on the premise that, in general, the more assets one possesses or has access to, the more positive the developmental outcomes.

There is little doubt that risks tend to increase vulnerability, or that assets can compensate for exposure to risk factors. The more one is protected from daily adversity the less one is adversely affected. It should be noted, however, that risks and assets are meaningful within the contexts of socially constructed models of what it means to develop positively.

EVOLVING CONCEPTS OF STRUCTURES

Having regularly emphasized structural dimensions of environments, Bronfenbrenner (1999), again addresses these issues, now providing more operational definition to the nature of environments. He continues to emphasize familiar concepts and themes including reciprocal interaction, structural ideas of nested ecological systems, and proximal process. He emphasizes regular activity with people, objects, and symbols, and that these activities are effective if increasingly complex. He also emphasizes time as a factor in the life course, based on a life course model (Clausen, 1986, 1993; Elder, 1974; 1998).

PSYCHOLOGICAL ECOLOGY AND GENERAL SYSTEMS THEORY

Bronfenbrenner's model is very popular. It is seen as a useful way to describe the structure and processes of individuals as they experience their environments. Following Bronfenbrenner's model, one might carry out an ecological analysis of person-environment interactions focusing on microsystems, mesosystems, exosystems, and macrosystems. Theoretically it is possible to superimpose this template on virtually any person-environment setting.

While it is possible to use a structural model as a metaphor for reality, it is nonetheless true that this model does not give a full account of developmental outcomes, even when one focuses also on proximal process. There is no question that persons interact with environments in complex ways. However the dimensions of this complexity yet remain to be adequately explained. As it stands, this model is, like many human ecological models, an oversimplification of an ecological account of human development.

Several questions enable one to assess psychological ecology in general. Answers to these questions lead toward general systems theory.

- If ecology is the study of ecosystems, to what extent does psychological ecology address ecosystem as the relevant unit of analysis?

- Does psychological ecology deal with the interaction of individuals with physical and chemical factors in the environment and exchanges of energy?

- Is the unit of analysis in psychological ecology maintained over time and as individuals come and go?

- Is there emphasis, beyond the level of the individual interaction, on patterns of interaction of a human biological community with environments?

Using Bronfenbrenner's model as representative of psychological ecology, it appears that the answers to these questions are not necessarily in the affirmative.

Structural Issues in Psychological Ecology

Psychological ecology, as developed by Bronfenbrenner, has rather consistently pursued a structural analysis of environments, in an effort to address the need to frame behavior and development within standardized contexts. At the same time, there has been little emphasis on general systems theory concepts, except for structural characteristics of ecosystems, such as are represented by microsystems, mesosystems, exosystems, and macrosystems.

Prior to the construction of these metaphors in psychological ecology, structural characteristics were represented in the social sciences in terms of culture, patterns of behavior, social class and mobility patterns, income, and occupation. Although there has been considerable recent interest in describing dynamic components, more specification of ecological systems is needed in psychological ecology.

Proximal Process and General Systems

The concept of proximal process is useful but does not account in specific terms for flows of energy and resources, as is the case from a general systems perspective. Possibly the concept of energy perturbations may be suggested or implied in terms of risks and dangerous environments. However, these terms do not provide a precise ecological account of action.

Regulation and Control of Systems

Psychological ecology is, in addition, undeveloped in terms of regulation and control of systems. Perhaps this a consequence of having devoted much attention to structures, such as microsystems and mesosystems, as well as more recent attention to system dynamics in terms of the proximal process.

Information Processing Components

Proximal processes appear to be an insufficient foundation for explaining information exchange. Certainly information is exchanged in proximal processes, but specifically how is this important in understanding human ecosystems? And how is information related to macrosystem or cultural concepts or to structural concepts, such as stratification and hierarchy?

While psychological ecology may recognize the importance of holism, interdependence, and interaction, it seems that is does so without sufficient attention to information components, and it remains rather limited in terms of explaining decision making processes and strategies used by individuals. Individual strategies may be described in terms of the utility of support systems and the consequences of support or its absence. However, precision seems limited by an undeveloped theory of information.

Disruptive Components

Disruptive components, another aspect of general systems theory, are abundant, yet limiting, in psychological ecology. There is much attention to such disruptions as violence, aggression, and destructiveness. However these phenomena are rarely seen in terms of more abstract and systems theory concepts, such as perturbations leading toward cessation of regular functions or system destruction, obstacles to reaching individual goals, obstacles to communication within a system, or forces outside a system that disrupt communication.

Life Process Components

Life process dimensions of general systems theory, including functions of living systems are seemingly implied or suggested. The language used in this regard tends to be about inputs and outputs, resources, and adaptation. However the focal point again seems to be largely on the individual rather than the composite, evolutionary change of the organization, processes of collective ecological adaptation, growth or decay of the system, termination of functions, or termination of the system.

Energy

Psychological ecology has not emphasized the flow of energy, production and consumption of resources - all essential ingredients in ecology. Because it has remained quite focused on psychological analysis, relying often on unique terminology, its taxonomies are not linked to natural science. Therefore, they are not particularly useful in building bridges to ecological science. Unique vocabularies tend to serve as obstacles to scientific communication.

Reductionism

The isolation of psychological ecology from ecological science by unique terms and concepts reinforces the reductionist tendency of psychological ecology to focus on the individual and on lowest common denominator structural concepts. This is in opposition to expanding a truly holistic framework of ecological inquiry.

Exercises
Chapter 4 CONCEPTS

There are many important concepts in this chapter on psychological ecology. In your opinion, what are the 5 most important concepts in this chapter? List and define them below.

1.

2.

3.

4.

5.

Chapter 4 EXAMINING THE CONCEPTS

Of all the concepts in this chapter, on psychological ecology, choose the three most important concepts. In your own words, explain why you believe they are the three most important concepts. Use comparisons and contrasts to explain your selection.

Chapter 4 APPLYING THE CONCEPTS

Two important concepts in this chapter on psychological ecology are microsystem and macrosystem. Choose two other important concepts from this chapter. For each of the two concepts you choose, write an example to show how it can be applied. Give the page number where your application would fit in the chapter.

1.

2.

Chapter 4 SYNTHESIS

This chapter on psychological ecology is divers in concepts and theoretical viewpoints. Your task is to write a final paragraph for the chapter. In this paragraph, bring together the most important thoughts in the chapter. Draw conclusions. Show how the important ideas in the chapter are related.

Chapter Five

Urban and Social Ecology

This chapter begins with foundational perspectives about cities and urban life and goes on to consider contemporary formulations of urban ecology and social ecology.

FOUNDATIONS OF URBAN ECOLOGY

Cities evolved and developed with changes in agrarian culture in the United States. As agricultural surpluses occurred, trade evolved. The development of cultures and cities based on trade, production, and consumption followed (Campbell, 1983).

One of the most well known foundations of urban social thought comes from the work of Tonnies (1940), who defined the concepts of Gemeinschaft and Gesellschaft. In the *Gemeinschaft*, relationships are based on traditions, solidarity, identity, and kinship. The Gemeinschaft is, essentially, a small town culture. In contrast is the urban center, which is known as *Gesellschaft*. Here one finds behavior resulting in isolation, competition, and loss of cohesiveness. In theory there is a circularity about these two systems. If Gemeinschaft seems to evolve into Gesellschaft, it appears that there are forces within the Gesellschaft that can influence change that will revert to a small town atmosphere – a return to Gemeinschaft.

A contemporary example has been termed the ***global village*** (McLuhan & Fiore, 1967; McLuhan & Powers, 1989), and in its realization in the form of the Internet in the 1990s. Some are now expressing concerns about how humans will adapt to an electronic global village. For example, in a global electronic ecosystem it is unclear how countless competing interests that are placed in direct contact will be reconciled, or what the impact will be on global values and standards.

If a multitude of face-to-face contacts in urban life results in segmentation, and superficiality, an electronic global ecosystem could amplify these characteristics. The problems associated with forces of control over behavior come immediately into focus. The fear is that individual control will be diminished as formal controls of processes of communication in the electronic ecosystem become increasingly restrictive.

BUILDING ON BASIC CONCEPTS

There were many other early sociological formulations of the experience of urban life. Durkheim (1933; 1986) for example, considered cities in terms of mechanical and organic solidarity. Perhaps his most well known concept is *anomie*. Durkheim described anomie in terms of the breakdown of social regulations, when society is no longer able to control individual behavior, resulting in relative normlessness. Durkheim conceived of normlessness as a condition of social structure.

Simmel (1917; 1971)examined the psychological characteristics of people in urban centers. He concluded that they had unique characteristics, such as indifference and apathy.

Wirth (1938) was among the first sociologists to consider the unique experience of the urban way of life. He defined the urban experience in terms of population size, density and heterogeneity, linking these aspects of the city to disorder, competition, and other adverse outcomes. Size and density are not the only salient characteristics of the urban environment. Cities feature other variables with which we have become very familiar (Lyon, 1989). Wirth framed his consideration of the city in terms of the city as a unit of analysis, involving interdependencies and competition (Smith, 1979). He also discussed the impact of the city on human relations, especially the characteristics of impersonality and superficiality that can be associated with urban life.

The Chicago School of urban sociology was developed by Robert Park (Park, 1915; 1923). The intellectual foundations of the University of Chicago, including John Dewey, George Herbert Mead, and Jane Addams, provided a background for the development of this unique form of social theory, which formed the foundation of urban ecology.

Park examined urban patterns and defined community in terms of territorially organized populations of mutually interdependent symbiotic individuals. Central in his view of ecology are basic ecological processes, such as distribution of organisms within environments, regulation of patterns of behavior for survival within niches, and division of labor.

Distribution of people in geographical areas refers to how residential, business and other areas are organized. These organizations are called *ecological units*, and ecological units form clusters of *ecological constellations*. Ecological units and ecological constellations are always changing, as people are mobile. Thus, social ecology is a study of dynamics of populations. These dynamics are complex, for as people move, they tend to concentrate and centralize.

Often groups within populations tend to segregate geographically in association with cultural, political, and economic forces. Groups compete for territory, with some taking over territories once occupied by others. These movements define the evolution of areas in terms of definable yet dynamic demographics and functional specialization. At times, population migration moves toward centralization, and at other times toward decentralization.

As one traces the evolution of urban ecology, one sees a focus on urban structure and design, as well as the functional aspects of cities. For example, Burgess (1926) and Burgess and Bogue (1964) provided descriptions of cities in terms of concentric zones. In his view, expansion involves individuals and groups. Zones were useful in describing patterns of land use and succession. Cities take on unique social, cultural and economic characteristics as their patterns of succession and zone formation evolve over time. Hoyt (1968; 1969) saw the design of cities in terms of sectors and devoted much attention to the practical aspects of land use planning for shopping centers and other civic facilities.

Subsequently Hawley (1950; 1953, 1975) developed another perspective on human ecology. His views were based on the concepts of Park and Burgess. Hawley focused on the community as the unit of analysis - the community in which populations adapt to their environments. He saw a certain tendency toward disorganization in human ecology, in particular the lack of congruence between biological ecology and various forms of human ecology.

DEVELOPMENTS IN URBAN ECOLOGY

In the early years of the 20th century, social scientists became interested in cities like Chicago as "urban laboratories". Much of the information gained from this early research was interpreted to suggest that people do not necessarily adapt well to the urban context, and maladaptive behaviors were found to be common.

Negative Aspects of Living in Cities

In the decades that followed, there was considerable work in the effects of evolving environments on individuals, families, and communities. For example, Milgram (1970) described the individual's experience of living in cities. Latane and Darley (1969) explained bystander apathy. Rainwater (1966) focused on fear. Zimbardo (1973) discussed human choice, individuation, reason, order and the lack thereof as functions of the city.

It has been noted often that, given the demographic characteristics of cities and the psychological overload that might be involved, people need to devote much time and effort to maintaining boundaries and proximity to, and interaction with, others. Moreover, it has become well known that large public housing facilities can provide experiences such as fear and lack of trust, while issues regarding social control, and monitoring of children can become central (Yancey, 1971).

Building on foundations of Wirth's analysis, Marsela (1998) identified an extensive list of urban variables thought to shape mental health and social deviancy and categorized these factors as environmental, sociological and economic, psychosocial, and psychological. Others have also investigated these factors (Bachrach, 1992; Ericksen, 1979; Glass & Singer, 1972; Goodhart & Zautra, 1984; Kasl & Harburg, 1975; Milgram, 1970; Neff, 1983; Raffestin & Lawrence, 1990; Rutter, 1981). These studies tend to focus on the psychology of urban life and how individuals adapt to cities (Marsella, 1998). A good deal of this work continues to focus on problems of living in urban areas, such as high population density, crowding, extreme heterogeneity, information overload, excessive stimulation, lack of communication, disorganized and antisocial behavior, crime, illegitimate births, divorce, and pathology.

Positive Aspects of Living in Cities

On the other hand, it should be noted that living in cities also has been described in more balance and positive terms. For example, Whyte (1980) challenged some of the long-held beliefs about effects of urban environments. He suggested, for example, that children may play in the streets because they prefer to do so - not necessarily because they lack playgrounds. According to Whyte (1980) cities have many attractive features, such as other people, comfortable places to sit, sun, and attractive housing arrangements.

CONTEMPORARY URBAN ECOLOGY

Urban ecology today serves as a cornerstone for urban design and planning, including natural and constructed environments, environmental health, economic and social justice. For example, Leitman (1999) describes urban issues as increasing population, pollution, "brownfields", eroding infrastructure, hazards, and other issues, for which there are definable solutions.

An emerging "New Urbanism" has a solutions-focused perspective, oriented toward less pollution, less waste, more human space, more human advantages, and less reliance on automobiles (Warren, 1998). In this perspective it is possible to respect the natural environment and establish effective human communities.

There is an emerging view of cities as human places (Breuste, Feldman, & Ohlmann, 1998), with problems of sustainability, environmental policy, environmental quality, urban vegetation, air pollution and air quality, heavy metals, and aesthetics. Urban life has been called "urban cosmology" (Grange, 1999), defined in part by normative consciousness, urban space patterns, symbolic perception, values, symbolism, meaning, and justice.

A renaissance of interest in cities is not unique to the United States. Karan and Stapleton (2000) examine contemporary Japanese cities such as Tokyo, focusing on urban planning, regulation, and government coordination. Ooi and Kwok (1997) examine urban life and design in Singapore, in particular the human constructed environment, town councils, the political economy, public housing, ethnic relations, crime, and security.

Perhaps in attempting to reduce the immensity of urban issues to manageable proportions, an interest in a holistic ecological analysis seems to have been lost. One is more likely to hear about risks, dangers, and fear in relation to the city. Fitzpartick and LaGory (2000) examine unhealthiness of urban areas, and a volume edited by Schell and Ulijaszek (1999) examines possibly deleterious urban effects on health.

Cities are made of the natural and the artificial. Thus some urban risks come from the natural environment, such as earthquakes, floods, and fires (Davis, 1998). Natural, social, and human constructed environments can clash in cities. Economics and technology are not always compatible with nature, and this is nowhere more evident than in cities. The theme of achieving a more effective integration of nature with technology in cities is threaded through many recent analyses of urban life (Levine & Upton, 1994; Platt, Rowntree & Muick, 1994).

URBAN LIFE AND HUMAN DEVELOPMENT

A relatively newly emerging area of research concerns the impacts of urban life on children. For example, Black and Krishnakumar (1998) suggest that urban life may be related to a number of outcomes, such as delinquency (Barone, Weissberg, Kasprow, & Voyce, 1995; violence (DuRant, Getts, Cadenhead, & Woods, 1995; and low educational expectations (Cook et al, 1996).

Discounting simple structural models, Black and Krishnakumar (1998) suggest that ecological models are useful in understanding the relationship of urban life on behavior and development. They also recommend ecological interventions, such as community involvement, intervention at multiple levels, and considering protective factors.

SUSTAINABLE URBAN ENVIRONMENTS

Recognizing that survival of cities as human places may be at stake, a new environmental imperative has focused on urban sustainability, equity, and justice (Roseland, 1997). Frey (1999) emphasizes restructuring cities toward sustainability, and Calthorpe (1993) discusses the benefits of moving, in the United States, from urban sprawl to ecologically sound communities. This would mean fewer automobiles, and more or other means of conveyance, including public transportation.

Achieving urban sustainability is a matter of values. Thus Golany (1995) sees the city as a complex project that can be improved by reassessing basic values, historical trends, and strategies of design. Such reassessment can help restore equilibrium as balance between humans and nature.

URBAN ECOLOGY IN TERMS OF HUMAN ECOLOGY

Contemporary urban ecology can be examined with reference to some of the characteristics of ecology, grounded in general systems theory, what we have previously discussed. In this regard it would appear that urban ecology remains, to a substantial extent, less a science of ecosystems than a study of issues and events, largely from disciplinary perspectives.

Thus we have the demography of urban social toxicity as research targets social class and poverty. We have the psychology of urban toxicity as we read an abundant and increasingly voluminous ethnographic and biographical literature of "the urban experience". Research that concentrates on selected urban experiences while controlling for others is fundamentally antiecological. Focusing, for example, on relationship of size of the city to youth violence while controlling for population density, in an effort to discover the "pure" effect of city size on violence specifically prevents understanding the complexity of the matter and mitigates against understanding how city size and population density, for example, might interact to have an effect on youth violence.

If the goal is to understand the ecological holistic complexity of urban life, it is counterproductive to define events as issues or problems. Thus we describe as limited, at best, urban research that has focused on specific aspects of cities, such as high rise housing, public places, and open spaces, in relation to human outcomes. Equally limited are studies of urban ethnicity alone, studies of social toxicity alone, urban socialization practices, urban mobility, or urban structures.

URBAN ECOLOGY AND GENERAL SYSTEMS THEORY

We have defined some of the essential components of ecological study: *structural*, *dynamic*, *governing*, *information processing*, *interrelationship*, *disruptive*, and *life process* components. In urban ecology, these ecological components must be understood as they are related to each other.

Structural Components

Studies of blocks, neighborhoods and other urban structures alone are limited in their implications for understanding cities and urban life. Urban structure is one essential component of an ecological analysis of cities, however structure must be considered in the context of other ecological factors.

Dynamic Components

Likewise, dynamic components associated with urban ecology include systems for distribution of resources and services, on the basis of which urban systems are maintained in varying degrees of stability. Above and beyond everyday issues of resource availability, there are special events that can have dramatic impact on urban life. These events can lead to changes in the balance of stability and disorder and seriously impact cohesiveness, adaptation, continuity, and discontinuity.

Governing Components

Known as governing components in general systems theory, these forces tend to be thought of as political, social, and psychological variables in urban ecology. In these terms, regulation in cities may be framed in terms of political power and wealth, rather than systems concepts of homeostasis, feedback, equilibrium, cybernetics, and individual and collective goals are generally not found in research on urban ecology.

Information Processing Components

Information processing components might focus primarily on urban residents' search for and use of information related to decision-making. How individuals process information is not generally an emphasis in urban ecology. It is more typical to find the emphasis on differential access to information and equity issues.

Interrelationship Components

Urban ecology can provide a framework for examining the urban area in terms of interdependence and holism, rather focusing on control of problems. For example, rather than dealing with violence as a symptom of city life, independent of other ecological phenomena, it is useful to examine political and economic dimensions of violence, as well as other conditions that might be related to violence, such as population migration, the pace of urban change, urban rebuilding, and major renovation.

Disruptive Components

Established urban routines change when there are dramatic changes in the built environment of cities as, for example, when during the 1960s massive urban renewal programs caused almost complete rebuilding of old cities, replacing them in some cases with new towns that were similar to strip malls. Thus patterns of individual and family behavior can be disrupted and new strategies of obtaining and exchanging resources must evolve.

Life Process Components

Urban ecology provides an opportunity to examine how production, consumption of resources, and the use and provision for storage of byproducts are inseparable from the lives of individuals in urban areas. In order to take full advantage of this opportunity it is crucial to take a completely holistic perspective on urban life rather than a perspective on one issue at a time. Thus, industries of production can be examined in systems related to systems of consumption, economics, social support, and family.

As it now stands, urban ecology tends to be an essentially issues-oriented field, based on independent perspectives of contributing disciplines. Demographers examine diversity and distribution of populations in urban areas, and associated issues, such as

ENVIRONMENTAL JUSTICE AND INJUSTICE

Environmental issues are central nowadays in virtually all aspects of the economy, politics, and culture. The issue is that forms of injustice are based on person-environment relationships. When resources are allocated differentially in association with group membership, or when there is inequitable distribution of environmental risk factors, there is environmental injustice and likely environmental racism. This can occur when policies are made, or actions are taken, concerning air pollution, water pollution, potential causes of climate change, distribution of industrial waste, destruction of natural areas and wetlands, consumption of pesticide residue, exposure of workers to toxic substances in the workplace, and a host of other equally deleterious events.

A Matter of Access

The issue in environmental justice is not simply whether the environment is being polluted. Nor is the issue whether growth is a good idea, for growth is inevitable and can be understood in terms of ecology - relationships of humans to their environments. The issue is much more complex and focuses on whether certain groups of people in our population have disproportionate access to more resources while becoming custodians and victims of disproportionate amounts of waste, pollution and deleterious toxic substances. This is environmental injustice.

Differential Exposure, Wealth, and Power

Environmental justice means not subjecting certain groups of people, often ethnic minorities and the poor, to systematic environmental disadvantage and risk. It means equal environmental rights. And it means concentrating not on the rights of the natural environment alone, but on the rights of all people. Problems for the natural environment are also problems for people, and vice versa. Thus pesticides in young children's environments may be problems for children's health, problems for their behavior, problems for school success, problems with regard to social behavior, and possibly problems for criminal justice.

Environmental injustice is a common experience of indigenous populations (Gedicks, 1993). They have been subjected to injustice from invasion during colonialization. When national parks were formed, native people lost enormous parcels of their resources. This is an expression of both environmental injustice and environmental racism.

Possession of capital and resources in our society tends to be directly related to possession of social power and inversely related one's exposure to pollutants (Odum & Odum, 1976). If the poor are exposed disproportionately to toxic substances and are therefore more likely to suffer illnesses, they will have to spend greater proportions of their incomes to getting well. They are less likely than the affluent to have health care insurance as well.

Environmental Justice Delayed

It seems there are always reasons to delay environmental justice. It is said that we need more data before taking action. Since much money and power are behind injustice, very high standards are set for proving that people are being harmed. It is argued that unless absolute proof can be established, no case can be made that the poor suffer any consequences from exposure to these substances. These standards may be criticized as delaying tactics.

Money is spent for the benefit of those who have money and power, because the poor have limited access to those who make decisions. Correcting the balance of power is fundamental if there is going to be environmental justice. If social ecology goals are to be accomplished, they will be accomplished through focused political action. Politically, this action seems to be grounded on the political left and in the Green Party. Political action that can effectively deal with these problems is likely to depend on the use of mechanisms that now exist. In particular, successful action will depend on grass-roots democratic action, including town meetings and committed citizen involvement.

Once people become deeply involved in local democratic decision-making, it is necessary to organize these democratic movements together in order to achieve larger scale social change. The power of the community is primary in bringing about local and national change. As collective decision-making for the good of the community transcends the interests of individuals, economic and social inequalities diminish.

To a certain extent an emphasis on human interest is contrary to - perhaps in balance with - action for the good of the environment. Social ecology seeks positive interaction of the human and natural ecosystems. Political and social institutions are at the interface between people and nature. Thus to understand social ecology, one needs to examine facets of issues relative to achieving a congruence between culture and nature.

At a level beyond social ecology there is a movement known as *radical ecology* (Merchant, 1992). Radical ecology tends to be more emphatic about the need for change and the degree of change needed. There is a tendency to go beyond ordinary environmentalism, broadening the scope of the problem to consideration of nonindustrialized nations and ecological issues and problems from a global perspective. It is not always possible to achieve public awareness of these perspectives if we are limited to bureaucratic and "within the system" solutions. While radical ecology is not necessarily revolutionary, it is useful to recognize that often the forces that exist within the system are opposed to change. Change can alter the balance of power.

A Matter of Social Justice

Social justice depends on achieving environmental justice. For example, for decades African Americans were not allowed to purchase homes in certain residential neighborhoods. They were kept out of these neighborhoods by a practice of realtors known as redlining. Systematic segregation of the poor and ethnic minority people has also been practiced by school districts in the form of maintaining essentially dual school systems for white and African American people.

In the 1960s and 1970s, the courts ordered metropolitan school desegregation plans to remedy this essentially inequitable practice. School segregation and desegregation are ecological events involving every aspect and institution of our society. The family, school systems, the legal system, including the courts, the U.S. Constitution and case law are all inescapably involved when schools are segregated. At the heart of segregation of people there is segregation of resources and power.

Although it is often alleged that in this separateness there is equality, from a legal perspective separate but equal is neither logical nor legal. Access to experiences, both good and bad, are inherently unequal. Separation of groups in our society is maintained by boundaries of social class, in neighborhoods, and in schools (Griffore, 1981; 1982; 1985). Where it occurs in schools, children may experience quite different learning climates.

SOCIAL ECOLOGY AND GENERAL SYSTEMS THEORY

As a science of ecosystems - human ecosystems, natural ecosystems and their relationships - social ecology shares some of the goals and the limitations of urban ecology. Yet it remains heavily focused as a study of issues and events. Because it tends to focus on ecosystems, research in social ecology is not customarily concentrated on one or two specific variables. Nor does social ecology research seek to be entirely quantitative in the framework of model building.

Social ecology seems to be generally congruent with components of general systems analysis: ***structural***, ***dynamic***, ***governing***, ***information processing***, ***interrelationship***, ***disruptive***, and *life process* components.

Structural Components

In social ecology, structures may be described in natural ecosystems and in social ecosystems. Over time there are incursions of human ecosystems into natural ecosystems and vice versa. In large part the extensive symbiosis of human and natural ecosystem is a synthesis of their structural aspects. This involves achieving acceptable patterns of human use of natural resources, less pollution of natural ecosystems with toxic chemicals, and more equitable distribution of resources and environmental risks to defined groups in human society.

Dynamic Components

Dynamic aspects of social ecology derive from the values and cultural sanctions associated with human use of natural ecosystems, and with the distribution of resources, services, and risks. Human motives that account for social segregation and environmental racism stand behind some of the dynamic forces at the heart of social ecology.

Governing Components

Of course at the core of social ecology are the forces of domination and hierarchy, which are so pervasive in human society but not in natural ecosystems. These forces are the foundation of action in human social life. They are governing components, from a perspective of general systems theory. Ecosystems are controlled by power, domination, and hierarchy.

Information Processing Components

Information processing is necessary for decision-making and allocation of resources and risks. Though having information can be useful in making decisions, perhaps a more important use of information comes in a critical science or liberating framework. In making decisions, people benefit from the knowledge of how decisions will affect their well being and their future. Shared access to information augments the probability of achieving social equity. This means eliminating differentiated access to information, and, in particular, eliminating differential access to accurate information.

Interrelationship Components

Social ecology specifically concentrates on interrelationship components. Examples are concepts of symbiosis, interdependence, and holism.

Disruptive Components

There are many forces that can disrupt the accommodation of natural and social ecosystems to each other. Lack of focus is one such factor. This can be exacerbated by preoccupation with frivolous and nonfocused belief systems and information, such as ecophilosophy and deep ecology. Social ecology avoids this to the extent it remains reality-based. Additional disruptive forces include forms of social action and movements that polarize and disunify people from each other and from natural esosystems. Any disruptive force that diminishes unity in diversity can be counterproductive relative the goals of social ecology.

Life Process Components

Like urban ecology, social ecology provides an opportunity to examine how production and consumption of resources and the use and provision for storage of byproducts are inseparable from the lives of individuals in urban areas and from their relationships with economic systems, systems of consumption, and family systems.

Exercises
Chapter 5 CONCEPTS

This chapter on urban and social ecology includes a wide variety of concepts. Which of these concepts are, in your opinion, the 5 most important concepts in this chapter? List and define them below.

1.

2.

3.

4.

5.

Chapter 5 EXAMINING THE CONCEPTS

Of all the concepts in this chapter, which are, in your view, the three most essential? In your own words, explain why you believe they are the three most important concepts. Use comparisons and contrasts to explain your rationale.

Chapter 5 APPLYING THE CONCEPTS

Choose two important concepts from this chapter on urban and social ecology. For each of these concepts, write an example to show how it can be applied. Give the page number where your application would fit in the chapter.

1.

2.

3.

Chapter 5 SYNTHESIS

This chapter has focused on history and contemporary information about urban and social ecology. Now that you have read this chapter, write a final paragraph in which you bring together the salient and essential thoughts and draw conclusions about this information from your perspective.

Chapter Six

Human Ecology in the Postmodern Era

What is postmodernism? It seems that there is no concise answer to this question. It has been said many times that postmodernism can mean anything or nothing. Gellner (1992) notes that it is a popular movement, but that it is not entirely clear exactly what it is. He doubts that there are postmodern essentials. He observes that postmodernism is represented in literature, philosophy, and anthropology. He notes that postmodernism seems to be antipositivistic and contrary to generalizations, but he is not sure about where the individual person stands with regard to postmodernism. He is certain that relativism is an essential characteristic of postmodernism.

A BRIEF SKETCH OF POSTMODERNISM

Despite many elements of uncertainty, some characteristics of postmodernism are clear. Postmodernism represents ideas that clash with modernity, which came upon the scene in the Renaissance and evolved together with the politics and economics of capitalism.

Sarup (1993) sees modernism as the culture of modernity and postmodernism is, he believes, the culture of postmodernity. It is against positivism, because positivism is grounded in the idea that there are facts. And it is against the idea that events can be explained. Facts are as they are constructed by an individual. There is in essence a denial of objectivism. Postmodernism incorporates a subjective point of view. It goes against the principle that people perceive the "truth". Thus one cannot assign much value to authoritative declarations of what the "truth" is. It is not supportive of inclusive descriptions of culture (Lyotard, 1984).

The Concept of Progress

Progress is a central concept in modernity. Progress is a simple concept. Its direction is clear, and the degree of progress attained is measurable. A postmodern view of reality reserves no space for a concept of progress, because progress requires some beginning point, a defined course, and a goal. In a postmodern view, it is not necessarily assumed that change is orderly and linear, based on identifiable forces of causality. How can there be progress when events unfold in capricious ways?

Progress and Developmental Change

Nor does a progress model apply to processes of personal development. In processes of individual development, there is no necessary relationship between events over time. Events are not necessarily caused by events that come before them, nor are they the causes of succeeding events. The flow of events is capricious and random, rather than ordered, linear, and determined.

Methodology and Postmodernism

In postmodernism, scientific method is inconsequential, and objective knowledge of facts is impossible. *Skeptical postmodernists* such as Baudrillard (1983) question the existence of a real world (Rosenau, 1992), and perceive reality as a collage of images and diverse truths achieved by consensus. Facts, as they are thought of in everyday language, do not exist as such. Derrida (1976, 1982) questioned meaning and fact, while placing emphasis on the method of deconstructing text. Foucault (1970), emphasizing power, also questioned durable truth and the factual basis of history.

Just as facts are rejected, so are theories, especially by skeptical postmodernists (Rosenau, 1992). To skeptical postmodernists, one theory is not necessarily more correct than other theories. Others, known as *affirmative postmodernists*, tend to be less rejecting of theories and are content to reform them (Rosenau, 1992).

Postmodern criticism goes beyond fact and theory to methodology. The search for new knowledge becomes difficult if one takes seriously the postmodern assertion that there is nothing new to be found, and that nothing can be known.

Using the technique of deconstruction of text, postmodernists seek the latent content and hierarchies reflected in written material. Postmodernists do not collect data from randomly selected samples, subject it to statistical analysis, and draw grand conclusions. They use narrative, interpretations, and ethnography to examine motivations and ideology that may be found in text.

Criticisms of Postmodernism

There are many criticisms of postmodernism. The rejection of scientific method is a point of alienation for positivistic scientists. Assertions that there is no objective reality, and that events unfold in an unpredictable sequence, are unacceptable for many as well.

For most people, it is comforting to believe in objective reality and in the predictability of events in everyday life. Indeed it is unsettling at best to think about a world in which virtually any event can follow any other event nondeterministically. Equally unsatisfying is the idea that rationality is of little use in a completely intersubjective world, where one cannot assess truth, and where any perspective is as defensible as any other view. If one were to accept fully these beliefs, one would need to reject the possibility of science as a foundation of scholarship in the social realm.

POSTMODERNISM AND HUMAN ECOLOGY

Bauman (1992) addresses postmodernism as it applies to social systems, observing that postmodernism is contrary to the assumptions of systems. Systems are generally considered to be cohesive, and consisting of elements that tend toward equilibration. As applied to social reality, however, a postmodern view is that reality is unstable and unpredictable. If one cannot predict successive states of a system, all possible states of a social habitat would therefore seem equally probable at any time (Bauman, 1992). Lack of certainty comes from the fact that the system is not guided by a single set of deterministic goals. There are multiple uncoordinated agencies in a complex system. They all act in the context of opportunities to access resources in order to address their unique adaptive problems. Since change is unpredictable over time, it is not useful to conceptualize development as a single and clear direction of change.

THE INDIVIDUAL IN POSTMODERNISM AND HUMAN ECOLOGY

Is the individual to be conceptualized as a passive recipient of the forces of environments within mechanical systems? Or, on the other hand, does the individual act within complex and unpredictable systems to construct a self, all the while challenging modern understandings of orderly, predictable, and generalizable patterns of developmental change? Perhaps Bronfenbrenner's principal accomplishment has been to make obvious this dilemma where modernity and postmodernity come together.

Underlying all of this is what Honneth (1992) characterizes as extreme individualization and a general posture toward the goal of individual self-construction. Honneth (1992) points out that the concept of individual freedom, especially with regard to the social environment, is defined in terms of being able to escape its controlling forces. Here is where there is a philosophical break between postmodernism from a sociological perspective and postmodernism from the perspective of human ecology. Ecologically speaking, interplay with environments is essential. Thus freedom from an ecological perspective can be defined in terms of how effectively an individual can relate to environments, changing adaptively within the habitats they provide.

The individual has consistently enjoyed the spotlight in contemporary life, especially since the advent of psychoanalysis. More recently, the decades of the 1970s and 1980s were known for their emphases on self and selfishness, respectively. Given this embeddedness of the self in social science, it is not entirely surprising that the individual seems to be at the center of attention in human ecology as well. Indeed a challenge one encounters when first getting acquainted with human ecology is that of making a distinction between a psychological understanding of events and an ecological perspective. The two points of view are clearly different.

It should be noted that in some branches of psychological theory, there is acknowledgment of, and even emphasis on, contexts of development. There are, for example, specific concepts of picking niches and creating environments (Scarr, 1992; Scarr & McCartney, 1983). At times, contexts are emphasized very heavily. Bronfenbrenner (1986), for example, has acknowledged the ease with which one can become oriented toward studying contexts more than persons. But in general, psychology has been about the individual.

The constant focus on the individual has caused human ecology to suffer from lack of clear identity and, thus, has tended to diminish its progress as science. To the extent that a lack of distinction exists between human ecology and psychology, there is a tendency to see human ecology as a reflection of popular culture and fashionable colloquialisms. Moreover, the importance of the individual has had another and quite different effect on human ecology. The field has become associated with a relatively small number of individuals who have gained prominence by shaping the field after their own ideas.

DEEP ECOLOGY: A CONSTRUCTION OF HUMAN ECOLOGY

Bronfenbrenner's (1979, 1989) human ecological perspective on human development is as an example of a particular construction of human ecology - a postmodern approach - that, for many, has become virtually synonymous with human ecology.

Bronfenbrenner initially placed much emphasis on the structure of environments and later recognized that such an approach tends to ignore the individual. We see, in a fashion not unusual in psychological theory building, how the preferences, errors, and corrective attempts of one theorist can shape decades of thought. In this case we have a transition from a construction of ecology emphasizing structure to a different construction of ecology with more emphasis placed on incompletely defined interpersonal processes. This allows the researcher to project a personal view on the construction, emphasizing those aspects of it that are convenient.

What is Deep Ecology?

Deep ecology serves as another example of a postmodern construction of human ecology. In deep ecology, humans are considered to live in a state of unity with nature, interacting in complex ways with the biotic and nonbiotic world. Deep ecology is based on the belief that there are special and deep relationships among all living beings and a deep relationship of all living beings with the Earth. Due to emphasis on ecosystems, deep ecology transcends the issues-orientation of psychological ecology. It does not attempt to capture attention by emphasizing metaphors such as "assets" and "social toxicity", which are popular concepts associated with psychological ecology.

The theme of sustainability, which is now emerging as a focal issue within environmental psychology, derives from deep ecology, because it is in the vital interest of humans to achieve sustainable environments. Deep ecology is concerned with fundamental common life processes that are deeply important and deeply rooted within the past, present, and future of humans and the Earth.

Deep ecology is a social construction of human ecology derived from the perspectives of Norwegian philosopher Naess (1973), who introduced the term *deep ecology* in order to describe a form of ecological thinking that is in contrast to an approach of *shallow ecology*. Subsequently Naess has been one of the most well known proponents of deep ecology (Naess, 1989; 1993a; 1993b), as have many others (Devall, 1988b; Devall & Sessions, 1985; Fox, 1984; Sessions, 1993; Zimmerman, 1993). The essential distinction between shallow ecology and deep ecology is basically as follows.

Shallow Ecology

Shallow ecology represents ecological thinking and action in the service of humans, specifically for the benefit of business, economic development, and commerce. Shallow ecology follows the principle that the environment exists for humans and is to be preserved as a means of meeting human needs.

In shallow ecology, the fundamental purposes are to foster the economics of production and to enable technological development, by supporting superficial approaches to managing resources. Superficial approaches include recycling, and corporate gifts to environmental agencies. Such approaches could conceivably have some impact on the environment, but they do not deal with fundamental changes in the ways in which humans relate to the natural environment. Within the shallow approach the goals are to make the environment cleaner, to make people less susceptible to environmental risks, and, thus, to foster the growth of societies as we now know them. Thus, there is a mixed message in shallow ecology: save the environment while fostering economic growth and production.

Deep: Not Shallow

Deep ecology is a long-term approach to ecology. The assumption is that in order to achieve sustainable environments for living beings, it is imperative to do more than simply serve the purposes of economic development. It is not acceptable to show symbolic acts of caring about the environment while keeping the eye on the prize of economic expansion.

In deep ecology the purpose is to achieve a long-range balance between humans and all living things, and between the biotic and nonbiotic worlds. The purpose is not to maximize the resources available to industries of production and commerce. Deep ecology requires careful and deep questioning of the assumptions and values of a society, along with basic transformations of virtually all aspects of society and culture. These changes reflect deep relationships of humans with their environments.

History of Deep Ecology

Deep ecology incorporates a particular social construction of aspects of biological ecology, including wildlife ecology and population biology, and based in part on the work of Aldo Leopold and John Muir in the United States. Muir was a famous explorer and naturalist and the first president of the Sierra Club. He was influential in establishing the conservation movement with his inspirational writings, in which he urged harmony with, and protection of, the natural environment. Leopold's book, *A Sand County Almanac* (1949), reflects his work as a famous conservationist and the founder of wildlife ecology. Leopold formed his concept of "land ethic" through his experiences as a young person, living in natural environments. As an adult, Leopold was employed by the U.S. Forest Service, and it was in the context of these experiences that he came to understand that the land was a living organism and the depth of the relationship of humans to the Earth. Leopold advocated the need to think "like a mountain", or, in other words, as if one has attained unity with nature.

Further construction of the environmental movement was accomplished some years later in the united States by Rachel Carson, who has come to symbolize the conservation movement in the United States. She began her career in ecology as writer of conservation booklets and popular magazine articles and, in the 1950s, became interested in, and wrote about, how young people can learn about nature. Seeing that nature and humans were inextricably related, she became alarmed about the use of pesticides and wrote ***Silent Spring*** (Carson, 1962), in which she was very critical of the use of pesticides. This book was a call to action for conservationists and ecologists.

Following Carson's work Paul Ehrlich sounded the alarm about the increasing growth of the world's population. In his book, ***The Population Bomb*** (Ehrlich, 1968) he heightened public awareness of the growth of population and its relation to environmental pollution. Ehrlich and his colleagues have been continuously involved in the environmental movement (Ehrlich & Ehrlich, 1991; Ehrlich, Ehrlich & Holdren, 1977). Barry Commoner, seeing the seriousness of the coming environmental crisis, also augmented the public consciousness by fostering the introduction of politics into the ecological movement (Commoner, 1967; 1971; 1977; 1979).

Other Foundations of Deep Ecology

Though grounded substantially in the conceptual structure of the environmental movement, foundations of deep ecology are also found in the "new physics", which introduced and included elements of Eastern metaphysics and religion (Capra, 1982; 1984).

Bohm (1981), for example, observed that quantum theory challenges the mechanistic view that everything can be reduced to smaller and smaller elements. Bohm emphasized holism, an indivisible universe, and causality that is not limited to simple linear paths. The fine balance and complex relationship of all elements of the universe have been described in terms of the "Butterfly Effect" (Gleick, 1987), which comes from chaos theory. The idea is that the movements of a butterfly's wings can ultimately cause events to occur at a place on Earth quite remote from these movements. This example is meant to illustrate how very small forces can have very significant remote effects. These influential ideas concerning complexity and unity provide a powerful additional foundation for deep ecology.

The Gaia Hypothesis

Consistent with a foundation in new physics, the *Gaia Hypothesis* (Lovelock, 1979; 1988) also provides a powerful basis for the construction of deep ecology as we now know it. Gaia refers to the notion that the Earth is a living system, of which humans are a part. The universe is alive (Lukas, 1980), intelligent, and completely connected. While parts of the earth appear to be separate, they are connected everywhere.

There is an implicate order of complete connectedness in the universe (Bohm, 1981). Thus, any notion of separateness is false (Talbot, 1986). This idea is especially useful in the construction of deep ecology, for if humans are to reduce their intrusiveness in the balance of nature, it is first necessary to understand that humans are a part of the Earth. To treat the Earth with respect is to treat the self with respect. To treat the Earth destructively is to destroy the self.

Deep Ecology from Literature and Philosophy

Additional aspects of the conceptual foundation of deep ecology come from literature and philosophy. This literary genre is exemplified by Pulitzer Prize-winning volume of poetry, *Turtle Island* (Snyder, 1974).

In philosophy, the themes of deep ecology are perhaps slightly less accessible. The philosophical foundations of deep ecology can be traced from the thinking of Spinoza, for example, in particular an emphasis on the active nature of humans and their relatedness with the world. The work of Heidegger is often regarded as congruent with deep ecology (Zimmerman, 1993). Naess, the Norwegian philosopher who is responsible for introducing the term ***deep ecology***, has written about it in a quite accessible way, relating it vividly to his life experience. Consistent with the concept of a constructed deep ecological perspective, in his book *Ecology, Community, and Lifestyle*, Naess (1989) describes ecophilosophy as coming from experience and intuition. Naess devoted decades of his life to an academic career in philosophy and ended it when he came to realize that he must devote attention to urgent environmental problems. His account reinforces and augments in this respect Leopold, whose work reflects a prominent role for personal experience as well.

The Individual in Deep Ecology

There is an aspect of deep ecology that reflects a psychological orientation and focus on the individual. Naess (1989) believes that living beings have a universal right to unfold a life, reach potential and form identity. What Naess refers to as *ecosophy* involves the self-realization of individuals through a process of wider identification with the Earth - a strong sense of intimacy and unity with nature.

The concept of ecological self has been developed as a central concept of deep ecology (Bragg, 1996; Thomashow, 1995). Humans have a personal and seamless relationship with the natural world, transcending mere coexistence. It is an emotionally engaging connection - a respectful and passionate interplay between living person and living universe.

Deep ecology is based on the premise that the world does not exist for humans Other living beings and the nonliving world have intrinsic value, and it is not up to humans to assign these values to them (Regan, 1981). Humans, therefore, share the right to form identity, but not at the expense of the Earth. Humans have no right to pursue action to meet human needs, regardless of consequences for the Earth. Humans have the responsibility of recognizing other animals as inherently valuable and treating them with respect. Humans have the responsibility to bring themselves in balance with the rest of the world, by managing the size of the human population and the proliferation of human constructed artifacts. Thus far, there has been little long-term success in doing so. As noted by Devall and Sessions (1985), this requires fundamental ideological change.

In short, one needs to reach a point of constructing Earth, humans, and human-Earth relationships consistent with the premises of deep ecology. In the past humans have related to the Earth from an anthropocentric perspective, (Brown, 1995). This has caused needless exploitation and tragic waste of natural resources. Thus, it is necessary to break away from a psychological tradition of conceptualizing self as separate from the Earth. Humans exist in unity with the Earth. The concept of person-Earth unity is very different from the concept of self as ego, to which psychology has devoted so much attention.

An Ecological Self

Ecopsychology as a relatively new perspective has accepted the premise that an individual begins life integrated with the external world and then becomes separated from it as the ego is formed (Roszak, Gomes, & Kanner, 1995). The role of environments is important, and the development of ego strength is based on rational control of person-environment relationships. Yet, psychological ego itself captures most of the attention while external environments are in the background.

This brings us to a particular aspect of the social construction of deep ecology that clearly distinguishes it from psychology. In deep ecology, the ecological self is a product of transactions of persons with environments (Naess, 1993a; Naess, 1993b; Sessions, 1993). To have an ecological self is to have identity with the entire Earth. In forming identity with the whole Earth, from mountains to bacteria, one loses the egocentric point of view and replaces it with an ecocentric perspective. In deep ecology the key to self-realization is through identifying with something greater than the self. This is a search for profound unity with nature. Thus deep ecology provides the conceptual means to understand a deep and wide identification process and subsumes it as well.

Beyond identification there is an emphasis in deep ecology on the need for human action. One kind of action that is congruent with deep ecology is living a simple life, which is best for the Earth. Another action has to do with population control. Naess has expressed the view that there are too many humans on earth. He has suggested that the capacity of Earth is 100 million in order to sustain certain desirable conditions (Devall & Sessions, 1985). This suggests that the problems of human coexistence with nature are due to population pressure. An action agenda based on deep ecology would also include using political action and public demonstrations.

Deep Ecology from a Collective Point of View

Deep ecology often seems to speak mostly to individuals, urging them to live lightly on the Earth and to accept a particular social construction of one's place on Earth. However, there is a collective side of deep ecology as well. As Naess (1989) has observed, deep ecology is also political. Yet, there is a sense in which in deep ecology there is less than full recognition of its political potential. Those who believe in the premises of deep ecology limit their actions to their own personal impact on the environment, and some tend not to see the potentials of deliberate political action.

Political action based on deep ecology can take the form of population control, resource management, and reducing environmental pollution. Managing one's own relationship with the Earth is necessary but not sufficient for global change. We are often advised to "Think globally and act locally." One can take action in terms of self-realization and in local politics in support of the ends of deep ecology. In doing so, one is in touch with the global community.

DEEP ECOLOGY: HAZARDS OF BEING POSTMODERN

Deep ecology has many commentators and critics (Bradford, 1989; Byers, 1992; Cheney, 1987; Devall, 1988a; Sylvan, 1985a; 1985b). For many of them, deep ecologists are naive, ineffectual, dropouts who have failed to acquire the personal and social skills needed for success in contemporary society.

Those who are critical of the essence of deep ecology, for example, the philosopher Bradford (1989), ask, "How deep is deep ecology?" He identifies rather glaring logical contradictions at the heart of deep ecology. For example, he points out that while deep ecologists hold that all living things have an equal right to exist, they fail to recognize that this is an example of simply another form of anthropocentrism: projecting human categorical systems on nature. Perhaps these contradictions should not be entirely unexpected, since the core of deep ecology stems more from ideology and personal construction than from logic.

Ecofeminists tend to have points of agreement as well as disagreement with some aspects of deep ecology (Cheney, 1987; Merchant, 1992; Zimmerman, 1987). Naess has been criticized for his use of the term "man". While some may see this as failure to use politically correct language, others see it as a basic lack of awareness of aspects of men's domination of women, much like human domination of Earth. Ecofeminists often take exception to being blamed for this and hold men primarily responsible for domination of Earth.

Another issue is that scholarship in deep ecology is part literature, part philosophy, part science, and largely critical. Although deep ecology scholarship can be in a particular instance quite molecular and quantitative, it is, in the long term and in the aggregate, quite global and holistic. Similar to general systems theory, deep ecology tends to be focused on the whole rather than parts or the sum of the parts. It tends to be ultimately antireductionistic, yet perhaps molecular at the level of means. Specificity is useful in achieving the general goals of deep ecology. However, in the last analysis, deep ecological insight depends on examining the global level of ecological organization. For example, one cannot understand the complexity of humans in their relationships with the natural world by focusing on specific toxic pollutants in the ground water in a particular census tract in a specific city.

DEEP ECOLOGY AND GENERAL SYSTEMS THEORY

As in many of the emerging fields of human ecology, unique and idiosyncratic terminology stands in the way of clarity in deep ecology. For deep ecologists, this is not necessarily perceived as a problem. It is important for deep ecologists to be able to describe natural environments and to have scientific evidence in support of imminent environmental disasters. On the other hand, failure to achieve congruence with the aims of contemporary science is not always, for deep ecology, a serious issue.

Structural Components

On the other hand, deep ecology has several anchoring points in general systems theory, and GST provides a framework in which deep ecology can be understood as ecology. For example, deep ecology recognizes structural components of ecosystems such as culture, ethnicity, and social class, subsystems, boundaries, levels of organization, and hierarchy of levels. These systems concepts are essential, for example, in analysis of populations and toxic pollution.

Dynamic Components

Deep ecology's treatment of dynamic components is also congruent with general systems theory. For example the flow of energy and its disruptions are key to understanding how changes in environments can have a range of consequences, some perhaps quite remote from the source of energy.

Governing Components

Deep ecology places governing components of systems at the heart of its focus, in particular issues relating to homeostasis in natural systems and human regulation and control of environments.

Information Processing Components

Information processing components are perhaps less evident in deep ecology. It is true that individuals need information as they make decisions about relating to natural environments in their daily lives, and that the process of self-realization requires information feedback to the self. Yet, the use of information relative to communication tends to be relatively limited in deep ecology to the extent that deep ecologists are not always overtly political.

Interrelationship Components

Perhaps most central in deep ecology are interrelationship components. For example, at the heart of deep ecology are characteristics of holism, the nature of interdependence and interactions of humans with nature, long-term processes and strategies of achieving optimal human-environment relationships, and aggregation of individual-environment interactions to the collective level.

Disruptive Components

Equally central to deep ecology are disruptive components. From the perspective of deep ecology, human behavior is the key disruption and cause of system decomposition. Perturbations caused by humans lead toward destruction and discontinuity, ultimately posing threats to the integrity of natural systems and to humans as well, ranging from physical and psychological stress to economic disruption. When natural systems are destroyed, there will be, at some point, serious degradation of human constructed and social systems as well.

Life Process Components

Thus, ultimately, life process components are also seriously and adversely affected, due to disruption of basic processes that sustain life: for example resource input and utilization, processes of adaptation, and reproduction. In the long term, affected as well are processes of evolutionary change, growth and/or decay of natural systems, and possible termination of life systems. According to the Gaia Hypothesis, the Earth is a living system. Thus, life process components apply to the sustainability of the Earth.

A MATTER OF SOCIAL CONSTRUCTION

Ultimately, from a postmodern perspective, all depends on how issues are framed and described. In psychological ecology, everything depends on what structures and interaction processes are called, and how they are described. In deep ecology, everything depends on how one constructs the natural environment and the self. As Gergen (1999) has pointed out, how things are described is not necessarily related to the way the world really is, but, rather to how the world is constructed, based on personal experience and relationships. We see the truth that we have been reinforced to see.

The field of human ecology presently features the constructions of a handful of scholars who seek to establish and defend their positions on what human ecology is about. As they construct their positions, they do not necessarily attend to competing constructions. Thus, human ecology consists of an often uneasy coexistence of multiple constructions. Constructions come and go, and there will always be new forms and ideas. Meanwhile, much effort goes into maintaining and discussing today's dominant constructions.

Is This Anything New?

The process of social construction and defense of construction seems to be an endless cycle. As Gergen (1999) points out, a multitude of different perspectives are possible, and no single perspective is necessarily accepted as representing the world as it really is. The fact is that, as Berger and Luckman (1966) observed, people collectively come to accept a particular view of reality as plausible.

In social constructionism, a central idea is that a person or event has meaning only in terms of context or culture (e.g., Gergen, 1985; Gergen & Davis, 1985; Sampson, 1983, 1985). From this perspective, the viewpoint of Erikson (1959) on life span development is, for example, a construction developed within the context of his experiences (Franz & White, 1985). From a constructionist point of view, his reality is simply fabricated on plausible grounds.

As one becomes enmeshed in a construction, it is often difficult to see it as something other than reality. The sense of reality is strengthened by social reinforcement in the form of recognition, fame, and subservience. As theorists construct the realities that become known as theoretical positions, their metaphors come to seem true.

Habitat and Sustainability

Bauman's (1992) characterization of postmodernity, especially a focus on habitat, is especially pertinent to human ecology. For example, he points out that the focus is on the habitat - consisting of all possible resources available relative to action - not simply on the agency. The habitat does not act in a determining fashion.

In human ecology the individual cannot be understood alone and isolated from environments. On the other hand, environments and individuals are highly interactive and mutually shaping and influencing, while the individual is left with the inevitability of decision-making and self-construction within defined environments.

The future of human ecology in the postmodern world depends on achieving a balance and a complete interaction between individuals and the habitat. It is this balance, or lack thereof, that is at the heart of the issue of sustainability. The issue to which human ecology can address itself is no less than the question of human survival, and the essence of the problem comes to sharp focus exactly at the intersection of the individual and the habitat.

The complex facets of this balance are the essence of recent work of Gardner & Stern (1996), Howard, (1997; 2000), Oskamp (2000), Stern (2000), and Winter (1996; 2000). Orr (1992) sees the coming of three crises. The first is a food crisis, which is interrelated with population growth. The second crisis comes with the end of cheap energy. The third crisis is caused by global climate change. Thus, plans for a sustainable society must deal with all three crises.

The Balance as a Problem of Behavior

Concern about the environment and sustainability has a history, especially in psychological ecology. Maloney and Ward (1973) and Maloney, Ward, and Braucht (1975) conceptualized the ecological crisis as a problem of maladaptive behavior. They developed a scale to measure ecological attitudes, originally including a verbal subscale and other subscales to measure commitment, affect, and knowledge. A revised scale, reported in 1975, was improved and shortened. Their research was prompted at least in part by their acknowledgment that ecological psychology was essentially about describing the effects of environments on people much more than the effects of people on environments.

Ecological problems are not simply problems of natural resources and technology. The ecological crisis is essentially caused by human behavior, and this is, it has been suggested, the business of psychology. Yet the road ahead will be difficult for several reasons. As Maloney and Ward (1973) have observed, people generally have rather high levels of verbal commitment and affect associated with the environment, but this is not necessarily accompanied by high levels of actual commitment or action, nor are they necessarily very knowledgeable about the environment.

People and Environments Interacting

Oskamp (2000) has stated the issue clearly. In his terms, the question is whether tomorrow's world will be worth living in. While this question suggests a focus on such problems as global warming; the greenhouse effect; loss of the ozone layer, forests and species; and exposure of people to toxic substances in the environment, it is quite evident that these are not the foundational problems.

The basis for all these problems may be framed in terms of human relationships to natural and constructed environments. Human behavior can reduce pollution, conserve resources, and regulate consumption. Human behavior can bring about a sustainable future or make it impossible. Oskamp (2000) suggests that matters of human behavior changes are within the province of psychology. Thus psychological ecology can take on a new focus – the technical basis for bringing about effective prevention and intervention in person-environment interactions.

Howard (2000) also has pointed out that several human lifestyles are associated with ecological problems that can threaten the quality of family life. He notes that changes in political and economic systems would be associated with changes in lifestyles that could diminish these ecological problems. Here again, psychologists can play a role in advancing the cause of a sustainable society.

These positions are intended to improve the interaction patterns between humans and nature. This could be a difficult task, because the patterns of human behavior that have become so ecologically destructive have been institutionalized as essential parts of culture. Thus massive system change is needed to de-emphasize basic premises, such as the need for continuous, unlimited growth, and confidence in technology to repair every damage done by humans to nature.

A Matter of Values

It is one thing to assert that very significant change in human behavior is imperative and quite another thing to bring about these changes. Actual change must be based on change of motivation and value structures. Oskamp (2000) identifies several possible techniques that can bring about behavior change. One technique is to live a simple lifestyle. This is sometimes referred to as voluntary simplicity and is a contrast to the premise that life is worth living only if one accumulates more and more possessions and if life becomes increasingly complex. This is also known as "living lightly on the earth" (Elgin, 1973). People are more likely to live simply and lightly if they have guidance about what this means and how it can be done.

One thing is clear. In order to live simply, people need to be supported by society's institutions. It is no good if people decide to live simply while the full force of technology drives society in the direction of complexity. Simple living should be aided by technological advances.

But perhaps simple living is too simple a solution to such a complex problem. For example, citing the work of Maloney and Ward a quarter of a century ago, Winter (2000) characterizes the complexity of human behavior in terms of depth psychology (object relations theory and ecopsychology), behavioral analysis, social psychology, and cognitive psychology. From a behavioral perspective she references Skinner's (1971) call for a technology of behavior. From social psychology she notes the importance of group dynamics and problem definition. From a cognitive perspective, she notes the relevance of how information is processed, and how errors can be made in information processing.

Echoing the theme of complexity, Stern (2000) describes the causes of behavior relative to environments as multidetermined ie. shaped by many factors including personal and contextual circumstances. Thus, while education can be useful, it cannot alone be expected to produce definitive outcomes relative to environmental behavior. Nor will programs intended to motivate correct action, perhaps based on fear, be necessarily successful. Indeed, by themselves, such measures might be counterproductive. Nor will incentives based on behavior control techniques or processes of behavior emulation or modeling alone necessarily be successful.

Though human ecology is the study of individual and environment relationships, a large part of the agenda that lies ahead for human ecology is about shaping organizational behavior as well. As Stern (2000) points out, while environmental problems such as pollution are caused by human behavior, it is often the collective behavior of organizations that can be more destructive than the behavior of individuals.

HUMAN ECOLOGY AND SCIENCE IN THE POSTMODERN ERA

Human ecology in the postmodern era remains grounded in positivistic science as it pertains to the relationships of humans and their natural environments. When looking at the relationships of humans with their social and human constructed environments, the questions of human ecology in the postmodern era are more readily answered by interpretive and critical modes of science. That is because the aims of the postmodern human ecology are not simply to understand in presumably generalizable and objective terms. From a postmodern viewpoint, The concern is not about quantitative mapping of results of linear causality.

In the postmodern era, there is a need to bridge positivistic science with postmodern perspectives. As we have suggested, general systems theory is a useful tool for grasping the complexity of an event. Thus we have described an approach to human ecology that rests on general systems theory. This approach resides in the modern world as well as in a postmodern era.

Since facts are considered to be constructions of individuals, human ecology in the postmodern era tends to emphasize *idiographic* science more than universal generalizations. In the postmodern era, there is recognition that it is not particularly useful to speculate about causes of each event, for they cannot be explained definitively, since knowledge is contextual, openly relativistic, and subjective.

Nor is it assumed that any particular presumed set of facts at any moment will necessarily lead to any other specific n future set of facts or scenario. Thus, human ecology in the postmodern era would do well to focus critical attention especially on the concept of progress. Change is capricious and can occur in any direction and any time. It is not necessarily an orderly of steps and transitions.

Ecology and Human Development in the Postmodern Era

The complexity of change in systems is mirrored in processes and outcomes of human development. As systems do not necessarily change in linear ways, development of individuals is not necessarily expected to be linear and predictable.

Human ecologists in the postmodern era do not exclusively base their scholarship on collecting data from randomly selected samples and then subjecting it to complex statistical analysis. They employ narrative, interpretations, and ethnography to examine motivations inherent in text that reveal ideology. They deconstruct presumed meanings of ecological events as they are constructed as socially acceptable consensus. Reflecting the foundations of postmodern methods, facts are questioned and deconstructed rather than regarded as universal.

In the postmodern era there is a rejection of dominant theoretical positions and idiosyncratic constructions, since they are, in fact, constructions of particular individuals. By employing the method of deconstruction it is possible to explore the latent material and motivations that are reflected in the text of human ecology theories. This can reveal the hierarchies of power, irrationalities, and unfounded assumptions on which theories are based.

In the postmodern era it is evident that human ecology must resolve that the unit of analysis is not the individual, but rather the individual in interaction with environments. Having resolved this issue, it is then necessary to go on to the postmodern corollary, which is that when persons interact with environments, meaning is constructed, subjective, relative, and contextual.

This postmodern corollary is crucial in understanding virtually any contemporary phenomenon or problem. The problem of achieving sustainable human environments, for example, brings to focus the importance of constructed meaning. Sustainable environments will likely not be achieved and maintained unless they are perceived as constructed and personally meaningful.

Exercises
Chapter 6 CONCEPTS

List and define 5 essential concepts found in this chapter on human ecology in a postmodern era.

1.

2.

3.

4.

5.

Chapter 6 EXAMINING THE CONCEPTS

From the list of concepts in the Chapter 6 Exercise on Knowledge, choose the three most important concepts or ideas in this chapter. Explain why you believe they are the three most important concepts or ideas. As you answer this question, use comparisons and contrasts to make your points.

Chapter 6 APPLYING THE CONCEPTS

From the concepts you defined in Chapter 6 Exercise on Knowledge, choose two important concepts. For each of these concepts, write an example to show how it can be applied. Give the page number where your application would fit in the chapter.

1.

2.

3.

Chapter 6 SYNTHESIS

Postmodernism incorporates many interesting concepts and ideas. Based on what you have read in this chapter, write this chapter's final paragraph. In this paragraph, bring together the chapter's essential ideas and draw conclusions about postmodernism and human ecology.

Ecological Exercises

Ecological Exercise #1

Think Ecology: Local and State

Locate and identify organizations and institutions in which you can get involved to work toward goals of ecological reclamation, renewal, and betterment

List and describe two *local* organizations.

1.

2.

List and describe at two organizations *your home state*.

1.

2.

Ecological Exercise #2

Be An Ecological Volunteer

What volunteer activities can you get involved in that might provide support to families and communities? Use the Internet, local directories, and other sources to identify volunteer opportunities. Focus on those that could help families and communities. Identify and describe *four possible opportunities*.

1.

2.

3.

4.

Ecological Exercise #3

Your Natural Environment

In almost every urban, suburban or rural area there are examples of the natural environment being destroyed and replaced by human constructed environments. Consider the area where you live. What are some examples of loss of natural environments and replacement with human constructed environments? List and describe three of these areas. Give the exact geographic location in terms of street or road address. Describe the previous natural condition of this area and describe how it is being replaced.

1.

2.

3.

Ecological Exercise #4

Toxic Substances in the Environment

Use the Web site of the U.S. Environmental Protection Agency's Enviromapper   http://www.epa.gov/enviro/html/em/index.html   to locate and describe toxic pollution in the community where you live. Use your own Zip Code as a geographic focal point. Identify three kinds of pollutants you find in your Zip Code area, and specifically as precisely as possible the addresses where you find them

1.

2.

3.

Ecological Exercise #5

Toxic Pollutants In Your Residence?

Go to a local hardware or other store where they have products for detecting pollutants in the home. Purchase at least one product, for example, one that can test for the presence of radon, or the presence of lead in the drinking water. Follow the product's directions carefully in your residence and report the results of your test.

1. What product did you purchase?

2. What substance is the product designed to test for?

3. Using this product, what did you find in your residence?

Ecological Exercise #6

An Ecosystem Analysis of Your Living Space

Assess the place where you live - it could be an apartment, home, or dorm room.

Space Survey:

Internal space: Begin by writing down all essential activities that take place within your living space such as eating, cooking, sleeping.... Then put these activities in priority order. Do you have sufficient space to meet all of your activities and needs?

Yes ____
No ____

If there is not enough internal space, what can you do to convert unused space or have more than one function to accommodate an already used space? Explain.

External space: The outdoor space can be analyzed using a similar procedure as above Prioritize all important activities and functions for you, such as going to the gym, eating out, etc. How far or close are these activities /functions to where you live?

	Activities/Functions	Walking distance	Driving distance
1.			
2.			
3.			
4.			
5.			
6.			

Lighting: How well is daylight used in your living space?

Appliances: Which appliances are essential to you? List these.

1.

2.

3.

4.

Energy Survey of Your Living Place:

List all the places where electricity is used.

How is your internal living space heated?

____gas ____electricity ____don't know

As you perceive your environment, check the overall efficiency of the heating arrangement in your living space.

____good ____fair ____poor

Water: What is the quality of your drinking water? (This is a subjective judgment as to quality)
____good ____fair ____poor

Sound: Check around your living space for sound proof qualities. Rate the overall sound proof quality of your living area.
_____good _____fair _____poor

Are there any weak points where sound travels easily through the walls, floors, ceilings or windows?
_____No _____Yes

If Yes, explain where:

Plants: Do you have locations inside where plants can survive?
_____No _____Yes

Do you have any plants?
_____No _____Yes

If Yes, how many plants do you have? _____.

Materials: Is the house made of durable materials in terms of integrity of the structure or is there potential for risk for damages such as water damage? Take a walk around the living place to check for cracks, such as those around your windows.

_____No problems _____Yes, potential problems\

If Yes, describe the problem.

Animals: Do you keep a pet in your living space?

_____No   _____Yes

People: Do others share your living space?

_____No _____Yes

	Age	Gender	Relationship to you
1.			
2.			
3.			
4.			

Bibliography

Agar, M. H. (1982). Toward an ethnographic language. *American Anthropologist, 84,* 780-795.

Allen, T.H. (1978). *New methods in social science research.* New York: Praeger.

Andrews, M., Bubolz, M., & Paolucci, B. (1980). An ecological approach to study of the family. *Marriage and Family Review, 3,* 29-49.

Ashby, W. (1958). *Introduction to cybernetics.* New York: Wiley.

Bachrach, L. (1992). The urban environment and mental health. *International Journal of Social Psychiatry, 8,* 5-15.

Bandura, A. (1977). *Social learning theory.* Englewood Cliffs, N.J.: Prentice-Hall.

Bandura, A. (1978). The self system in reciprocal determinism. *American Psychologist, 33,* 344-358.

Bandura, A. (1982). Self-efficacy mechanism in human agency. *American Psychologist, 37,* 122-147.

Bandura, A. (1986), *Social foundations of thought and action: A social cognitive theory.* Englewood Cliffs, NJ: Prentice-Hall.

Bandura, A. (1988). Perceived self-efficacy: Exercise of control through self-belief. In J. P. Dauwalder, M. Perrez, & V. Hobi (Eds.), *Annual series of European research in behavior therapy* (Vol. 2, pp. 27-59). Amsterdam/Lisse: Swets & Zeitlinger

Bandura, A. (1992). Exercise of personal agency through the self-efficacy mechanism. In R. Schwarzer (Ed.) *Self-efficacy: Thought control of action* (pp.3-38). Washington, DC: Hemisphere.

Bandura, A. (1993). Perceived self-efficacy in cognitive development and functioning. *Educational Psychologist, 28,* 117-148.

Bandura, A. (1995). Exercise of personal and collective efficacy in changing societies. In A. Bandura (Ed.) *Self-efficacy in changing societies* (pp. 1-45). Cambridge: Cambridge University Press.

Barker, R. (1968). *Ecological psychology*. Stanford: Stanford University Press.

Barker, R. & Wright, H. (1951). *One boy's day*. New York: Harper.

Barone, C., Weissberg, R. P., Kasprow, W., & Voyce, C. K. (1995). Involvement in multiple problems behaviors of young urban adolescents. *Journal of Primary Prevention*, 15, 261-283.

Bateson, G. (1972). *Steps to an ecology of mind*. New York: Ballantine.

Bateson, G. (1979). *Mind and nature*. New York: Dutton.

Baudrillard, J. (1983). *In the shadow of the silent majorities*. New York: Semiotext.

Bauman, Z. (1992). A sociological theory of postmodernism. In P. Beilharz, G. Robinson, & J. Rundell (Eds.). *Between totalitarianism and postmodernity* pp. 149-162. Cambridge, Massachusetts: The MIT Press.

Bennett, J. W. (1976). *The ecological transition: Cultural anthropology and human adaptation*. New York: Pergamon Press.

Berger, P. & Luckmann, T. (1966). *The social construction of reality*. New YorK: Anchor Books.

Bertalanffy, L. von. (1968). *General system theory*. New York: George Braziller.

Black, M. M. & Krishnakumar, A. (1998). Children in low-income, urban settings. *American Psychologist*, 53, 635-646.

Block, J., Block, J. H., & Keyes, S. (1988). Longitudinally foretelling drug usage in adolescence: Early childhood personality and environmental precursors..*Child Development*, 59, 336-355.

Bloom, B. S. (Ed.) (1956) *Taxonomy of educational objectives: The classification of educational goals: Handbook I, cognitive domain*. New York ; Toronto: Longmans, Green.

Bogenschneider, K. (1996). An ecological risk protective theory for building prevention programs, policies, and community capacity to support youth. *Family Relations*, 45, 127-138.

Bohm, D. (1981). *Wholeness and the implicate order*. London: Routledge & Kegan Paul.

Bookchin, M. (1962). *Our synthetic environment*. New York: Harper and Row.

Bookchin, M. (1965). *Crisis in our cities*. Englewood Cliffs, N.J.: Prentice-Hall

Bookchin, M. (1980). *Toward an ecological society*. Montreal: Black Rose Books

Bookchin, M. (1982). *The ecology of freedom*. Palo Alto, CA: Cheshire Books.

Bookchin, M. (1986). *The modern crisis*. Philadelphia, PA: New Society Publishers.

Bookchin, M. (1988). Social ecology versus deep ecology. *Socialist Review*, 18, 9-29.

Bookchin, M. (1990). *Remaking society: Pathways to a green society*. Boston: South End Press.

Bookchin, M. (1996). *The philosophy of social ecology*. 2nd edition. Montreal: Black Rose Books.

Boss, P. (1987). Family stress. In M. B. Sussman & S. K. Steinmetz (Eds.) *Handbook of Marriage and the Family* (pp. 695-723). New York: Plenum.

Boyden, (1986). An integrative perspective to the study of human ecology. In R. J. Borden, J. Jacobs, & G. L. Young (Eds.) Human ecology: A gathering of perspectives. (pp. 3- 25). College Park, Maryland: The Society for Human Ecology.

Bradford, G. (1989). *How deep is deep ecology?* Hadley, Massachusetts, Ojai, California: Times Change Press.

Bragg, E. (1996): Towards ecological self: Deep ecology meets constructionist self theory. *Journal of Environmental Psychology*, 16, 93-108.

Breuste, J., Feldman, H., & Ohlmann, O. (1998). *Urban ecology*. Berlin: Springer-Verlag.

Bristor, M. (1990). *Individuals, families, and environments*. Dubuque, IA: Kendall/Hunt.

Broderick, C. & Smith, J. (1979). The general systems approach to the family. In W. Burr, R. Hill, R.I. Nye, & I. Reiss (Eds.), *Contemporary theories about the family: Vol. 2. General theories/theoretical orientations* (pp. 112-129). New York: Free Press.

Bronfenbrenner, U. (1979). *The ecology of human development.* Cambridge, MA Harvard University Press.

Bronfenbrenner, U. (1986). Recent advances in research on human development. In R. K. Silbereisen, K. Eyferth, & G. Rudinger (Eds.), *Development as action in context: Problem behavior and normal youth development* (pp. 287-309). Heidleberg and New York: Springer-Verlag.

Bronfenbrenner, U. (1989). Ecological systems theory. In R. Vasta (Ed.), *Annals of child development,* Volume 6.(pp. 197-249). Greenwich, CT: JAI Press.

Bronfenbrenner, U. (1993). The ecology of cognitive development: Research models and fugitive findings. In R. H. Wozniak & K. Fischer (Eds.). *Scientific environments* (pp. 3-44). Hillsdale, NJ: Erlbaum.

Bronfenbrenner, U. (1999). Environments in developmental perspective: Theoretical operational models. In S. L. Friedman & T. Wachs (Eds.) Measuring environment across the life span, pp. 3-28. Washington, D.C.: The American Psychological Association.

Bronfenbrenner, U., & Morris, P. (1997). The ecology of developmental processes. In W. Damon (Ed.),*Handbook of child psychology*, pp. 993-1028. New York: Wiley.

Brook, J. S., Brook, D. W., Whiteman, M., Gordon, A. S., & Cohen, P. (1990). The psychosocial etiology of adolescent drug use: A family interactional approach. *Genetic, Social, and General Psychology Monographs, 116*(Whole).

Brook, J. S., & Newcomb, M. (1995). Childhood aggression and unconventionality: Impact on later academic achievement, drug use, and workforce involvement. *The Journal of Genetic Psychology, 156,* 393-410.

Brook, J. S., Whiteman, M., Gordon, A. S., & Cohen, P. (1986). Some models and mechanisms for explaining the impact of maternal and adolescent characteristics on adolescent stage of drug use. *Developmental Psychology, 22,* 460-467.

Brown, C. S. (1995): Anthropocentrism and ecocentrism: The quest for a new worldview. *Midwest Quarterly, 36,* 191-202.

Brown, M. (1993). *Philosophical studies in home economics: Basic ideas by which home economists understand themselves.* East Lansing: Michigan State University.

Bruhn, J. G. (1974). Human ecology: A unifying science. *Human Ecology, 2,* 105-125.

Bubolz, M., Eicher, J., & Sontag, M. S. (1979). The human ecosystem: A model. *Journal of Home Economics, 71,* 28-31.

Bubolz, M. & Sontag, M. S. (1993). Human ecology theory. In P. Boss, W. Doherty, R. LaRossa, W. Schumm, & S. Steinmetz (Eds.) *Sourcebook of family theories and methods: A contextual approach.* Pp. 419-448. New York: Plenum.

Buckley, W. (1967). *Sociology and modern systems theory.* Englewood Cliffs, N.J.: Prentice-Hall.

Burgess, E. W. (Ed.) (1926). *The urban community : Selected papers from the Proceedings of the American sociological society, 1925.* Chicago: University of Chicago Press.

Burgess, E. W. & Bogue, D. J, (Eds.) (1964). *Contributions to urban sociology.* Chicago: University of Chicago Press.

Byers, B. A. (1992) Deep ecology and its critics: A Buddhist perspective. *The Trumpeter, 9,* 33-35.

Campbell, B. (1983). *Human ecology.* New York: Aldine.

Capra, F. (1982). *The turning point.* New York: Simon and Schuster.

Capra, F. (1984). *The Tao of physics.* 2nd ed. New York: Bantam Books.

Capra, F. (1994). *Systems theory and the new paradigm.* In C. Merchant (Ed.). Ecology. (pp. 334-341). Atlantic Highlands, NJ: Humanities Press.

Calthorpe, P. (1993). *The next American metropolis: Ecology, community, and the American dream.* New York: Princeton Architectural Press.

Carson, R. (1962). *Silent spring.* Boston: Houghton and Mifflin.

Cheney, J. (1987) Eco-feminism and deep ecology. *Environmental Ethics, 9,* 115-145.

Chopra, D. (1989). Quantum healing: *Exploring the frontiers of mind/body medicine*. New York: Bantam Books.

Chopra, D. (1993). *Ageless body, timeless mind: The quantum alternative to growing old.* New York: Harmony Books.

Clark, J. (1998). Municipal dreams: A social ecological critique of Bookchin's politics. In A. Light (Ed.) *Social ecology after Bookchin*, pp. 137-191. New York: The Guilford Press.

Clarke, R. (1973). *Ellen Swallow: The woman who founded ecology*. Chicago, IL Follett Publishing Company.

Clausen, J. A. (1986). *The life course*. Englewood Cliffs, NJ: Prentice Hall.

Clausen, J. A. (1993). *American lives: Looking back at the children of the Great Depression*. New York: Free Press.

Cook, T. D., Church, M. B., Ajanaku, S., Shadish, W.R. Jr., Kim, J. R., & Cohen, R. (1996). The development of occupational aspirations and expectations among inner-city boys. *Child Development, 67*, 3368-3385.

Commoner, B. (1967). *Science and survival*. New York: Viking Press.

Commoner, B. (1971). *The closing circle: Nature, man, and technology*. New York, Knopf.

Commoner, B. (1977). *The poverty of power: Energy and the economic crisis.* New York and Toronto: Bantam Books.

Commoner, B. (1979). *The politics of energy*. New York: Knopf.

Darling, F. F. (1955-1956). The ecology of man. *American Scholar, 25*, 38-46.

Coontz, S. (1995). The way we weren't: The myth and reality of the "traditional" family. *Phi Kappa Phi Journal, 75*, 11-14.

Davis, M. (1998). *Ecology of fear: Los Angeles and the imagination of disaster*. New York: Metropolitan Books.

Derrida, J. (1976). *Of grammatology*. Translated G. Spivak. Baltimore, Maryland: Johns Hopkins University Press.

Derrida, J. (1982). Sending: On representation. *Social Research, 49,* 294-326.

Devall, B.(1988a). Deep ecology and its critics. *The Trumpeter, 5,* 55-60.

Devall, B.(1988b). *Simple in means, rich in ends. Practicing deep ecology.* Salt Lake City: Peregrine Smith Books.

Devall, B., & Sessions, G. (1984). The development of natural resources and the integrity of nature. *Environmental Ethics, 6,* 293-322.

Devall, B., & Sessions, G. (1985). *Deep ecology: Living as if nature mattered.* Salt Lake City Utah: Peregrine Smith.

De Young, R. (1999). Environmental psychology. In D. E. Alexander and R. W. Fairbridge (Eds.) *Encyclopedia of environmental science.* Hingham, MA: Kluwer Academic Publishers.

DuRant, R. H., Getts, A. G., Cadenhead, C., & Woods, E. R. (1995). The association between weapon-carrying and the use of violence among adolescents living in and around public housing. *Journal of Adolescence, 18,* 579-592.

Durkheim, E. (1933). *The division of labor in society.* Translated by G. Simpson. Glencoe, Ill.: Free Press.

Durkheim, E. (1986). *Durkheim on politics and the state.* Edited with an introduction by A. Giddens. Translated by W. D. Halls. Stanford, Calif. : Stanford University Press.

Ehrlich. P. R. (1968). *The population bomb.* New York: Ballantine Books.

Ehrlich, P. R., & Ehrlich, A. H. (1991). *Healing the planet: Strategies for resolving the environmental crisis.* Reading, Mass.: Addison-Wesley.

Ehrlich, P. R., Ehrlich, A. H., & Holdren, J. P. (1977). *Ecoscience : Population, resources, environment.* San Francisco : W. H. Freeman.

Elder, G. H., Jr. (1974). *Children of the Great Depression.* Chicago: University of Chicago Press.

Elder, G. H., Jr. (1998). *Children of the Great Depression (25th Anniversary Edition).* Boulder, Colorado: Westview Press.

Elgin, D. (1973). *Voluntary simplicity: Toward a way of life that is outwardly simple, inwardly rich* (Rev. Ed.). New York: Quill.

Ericksen, E. (1979). *The urban experience.* Austin, Texas: University of Texas Press.

Erikson, E. H. (1959). *Identity and the life cycle.* New York: International Universities Press.

Ferguson, M. (1980). *The aquarian conspiracy.* Los Angeles : J. P. Tarcher; New York: Distributed by St. Martin's Press.

Fitzpatrick, K. M. & LaGory, M.. (2000). *Unhealthy places: The ecology of risk in the urban landscape.* New York: Routledge.

Foucault, M. (1970). *The order of things: An archaeology of the human sciences.* New York: Pantheon.

Fox, W. (1984): Deep ecology: A new philosophy of our time? *The Ecologist, 14,* 194-200.

Frankl, V. (1962). *Man's search for meaning: An introduction to logotherapy.* Boston: Beacon Press.

Frankl, V. (1997). *Man's search for ultimate meaning.* New York: Insight Books.

Franz, C. W. & White. K. M. (1985). Individuation and attachment in personality development: Extending Erikson's theory. *Journal of Personality, 53,* 224-256.

Freud, S. (1953). *A general introduction to psychoanalysis.* Garden City, NY: Doubleday.

Freud, S. (1960). *The ego and the id.* New York: Norton.

Freud, S. (1961). *Civilization and its discontents.* New York: Norton.

Freud, S. (1965). *New introductory lectures.* New York: Norton.

Frey, H. W. (1999). *Designing the city: Towards a more sustainable urban form.* London; New York: E & F N Spon.

Gallagher, W. (1933). *The power of place.* New York: Poseidon Press.

Gardner, G. T., & Stern, P. C. (1996). *Environmental problems and human behavior*. Needham Heights, MA: Allyn & Bacon.

Gedicks, A. (1993). *The new resource wars*. Boston: South End Press.

Gellner, E. (1992). *Postmodernism, reason and religion*. New York: Routledge.

Gergen, K. J. (1985). The social constructionist movement in modern psychology. *American Psychologist, 40*, 266-273.

Gergen, K. J. (1999) *An invitation to social construction*. London: Sage Publications.

Gergen, K. J. & Davis, K. E. (1985). *The social construction of the person*. New York: Springer-Verlag.

Gibson, E. (1997). An ecological psychologist's prolegomena for perceptual development: A functional approach. In C. Den-Read & P. Zukow-Goldring (Ed.), *Evolving explanations of development*. Washington, D.C.: American Psychological Association.

Glass, D. & Singer, J. (1972). *Urban stress: Experiments on noise and social stressors*. San Diego, CA: Academic Press.

Gleick, J. (1987). *Chaos*. New York: Penguin Books.

Good, C. V. (1973). *Dictionary of education*. 3rd edition. New York: McGraw-Hill.

Golany, G. (1995). *Ethics and urban design: Culture, form, and environment*. New York: John Wiley & Sons.

Goodhart, D., & Zautra, A. (1984). Assessing quality of life in the community An ecological approach. In W. O'Connor & W. Lubin (Eds.) *Ecological approaches to clinical and community psychology* (pp. 251-290). New York: Wiley.

Gove, P. B. (1964). *Webster's third new international dictionary*. Springfield, Massachusetts: G.C. Merriam Co.

Grange, J. (1999). *The city: An urban cosmology*. Albany: State University of New York Press.

Griffore, R. J. (1981). Third generation school desegregation issues: An agenda for the future. In Procedures and policy research to develop an agenda for desegregation studies (pp. 149-227). A final report to the U.S. Department of Education. East Lansing, MI: Michigan State University.

Griffore, R. J. (1982). Effects of school desegregation in New Castle County, Delaware on family life: A preliminary study. In R.L. Green et al. Metropolitan school desegregation in New Castle County, Delaware. A final report to the Rockefeller Foundation (pp. 473-485). East Lansing, MI: Urban Affairs Programs, Michigan State University.

Griffore, R. J. (1985). School desegregation: Some ecological and research issues. In R. L. Green (Ed.) *Metropolitan desegregation* pp. 210-229. New York: Plenum Press.

Gross, M. L.(1978). *The psychological society*. New York: Random House.

Hartmann, (1991). *Boundaries of the mind*. New York: Basic Books.

Hawkins J. D., Catalano J. F., & Miller J. Y. (1992). Risk and protective factors for alcohol and other drug problems in adolescence and early adulthood: Implications for substance abuse prevention. *Psychological Bulletin, 112*, 64-105.

Hawley, A. H. (1944). Ecology and human ecology. *Social Forces, 22*, 399-405.

Hawley, A. (1950). *Human ecology: A theory of community structure*. New York: Ronald Press Co.

Hawley, A. (1953). *Intrastate migration in Michigan, 1935-1940.* Ann Arbor: University of Michigan Press.

Hawley, A. (Ed.). (1975). *Man and environment* . New York: New Viewpoints.

Hayward, T. (1994). *Ecological thought*. Cambridge: Polity Press.

Honneth, A. (1992). Pluralization and recognition: On the self-misunderstanding of postmodern social theorists. In P. Beilharz, G. Robinson, & J. Rundell (Eds.). *Between totalitarianism and postmodernity* pp. 163-172. Cambridge, Massachusetts: The MIT Press.

Hook, N., & Paolucci, B. (1970). The family as an ecoststem. *Journal of Home Economics, 62*, 315-318.

Howard, G.S. (1997). *Ecological psychology: Creating a more earth-friendly human nature.* Notre Dame, IN: University of Notre Dame Press.

Howard, G.S. (2000). Adapting lifestyles for the 21st century. *American Psychologist, 55,* 509-515.

Hoyt, H. (1968). *The changing principles of land economics. From horse and buggy to the nuclear age; fifteen 20th century revolutions that have changed land economics.* Washington: Urban Land Institute.

Hoyt, H. (1969). *The location of additional retail stores in the United States in the last one-third of the twentieth century; a research monograph for National Retail Merchants Association.* New York, National Retail Merchants Association.

Jung, C. J. (1954). *The development of personality.* Princeton: Princeton University Press.

Jungen, B. (1986). Integration of knowledge in human ecology. In R. Borden (Ed.) *Human ecology: A gathering of perspectives.* College Park, Maryland: Society for Human Ecology.

Kantor, D. & Lehr, W. (1975). *Inside the family.* San Francisco, California: Jossey-Bass, Inc.

Karan, P. P. & Stapleton, K. (Eds.)(2000). *Japanese city.* Lexington, Kentucky: University Press of Kentucky.

Kasl, S., & Harburg, E. (1975). Mental health and the urban environment: Some doubts and second thoughts. *Journal of Health and Social Behavior, 16,* 268-279.

Klein, J. T. (1990). *Interdiscplinarity: History, theory, and practice.* Detroit: Wayne State University Press.

Kockelmans, J. (1979). Why interdisciplinarity? In J. Kockelmans (Ed.). (pp. 123-160). *Interdisciplinarity and higher education.* University Park: The Pennsylvania State University Press.

Koestler, A. (1964). *The act of creation.* New York: Dell.

Kovel, J. (1998). Negating Bookchin. In A. Light (Ed.) *Social ecology after Bookchin*, (pp. 27-57). New York: Guilford Press.

Kuhn, A. (1974). *The logic of social systems.* Homewood, Illinois: Dorsey.

Kuhn, A. (1975). *Unified social science.* Homewood, Illinois: Dorsey.

Laszlo, E. (1972). *Introduction to systems philosophy.* New York: Gordon and Breach.

Latane, B., & Darley, J. M. (1969). Bystander apathy. *American Scientist, 57,* 244-268.

Leitman, J. (1999). *Sustaining cities: Environmental planning and management in urban design.* New York: McGraw-Hill.

Lyman, S. M., & Scott, M. B. (1975). *The drama of social reality.* New York: Oxford University Press.

Leopold, A. (1949). *A Sand County almanac.* Oxford: Oxford University Press

Le Vine, D. G. & Upton, A. C. (Eds.) (1994). *The city as a human environment.* Wesport, Connecticut: Praeger.

Lewin, K. (1935). *A dynamic theory of personality.* New York: McGraw-Hill.

Light, A. (1998a). Introduction: Bookchin as/and social ecology. In A. Light (Ed.) *Social ecology after Bookchin*, (pp. 1-23). New York: Guilford Press.

Light, A. (1998b). Reconsidering Bookchin and Marcuse as environmental materialists: Toward an evolving social ecology. In A. Light (Ed.) *Social ecology after Bookchin.* (pp. 343-383). New York: Guilford Press.

Lilienfeld, R. (1978). *The rise of systems theory: An ideological analysis.* New York: John Wiley and Sons.

Lovelock, J.(1979). *Gaia. A new look at life on the Earth.* Oxford: Oxford University Press.

Lovelock, J. (1988). *The ages of Gaia. A biography of our living Earth.* Oxford: Oxford University Press.

Lukas, M. (1980). The world according to Ilya Prigogine. *Quest/80, 4,* 86.

Lyon, L. (1989). *The community in urban society.* Lexington, MA:.D.C. Heath & Co.

Lyotard, J. (1984). *The postmodern condition: A report on knowledge.* Manchester: Manchester University Press.

Maloney, M. P. & Ward, M. P.(1973). Ecology: Let's hear from the people. *American Psychologist, 28,* 583-586.

Maloney, M. P., Ward, M. P., & Braucht, G. N. (1975). A revised scale for the measurement of ecological attitudes and knowledge. *American Psychologist, 30,* 787-790.

Marsella, A. J.(1998). Urbanization, mental health, and social deviancy. *American Psychologist, 53,* 624-634.

Martinez, E. (1988). Mexican American/Chicano families: Challenging the stereotypes. In H. B. Williams (Ed.), *Empowerment through difference: Multicultural awareness in education.* Home Economics Teacher Education Yearbook 8/1988. Washington, D.C.: Teacher Education Section, American Home Economics Association.

Maruyama, M. (1963). The second cybernetics deviation-amplifying mutual causal process. *American Scientist, 51,* 164-179.

Maslow, A.H.. (1954). *Motivation and personality.* New York: Harper.

Maslow, A. H. (1968). *Toward a psychology of being.* Princeton: Von Nostrand.

Maslow, A. H. (1975). *Motivation and personality.* New York: Harper.

McCall, R. B. (1990). Promoting interdisciplinary and faculty-service-provider relations. *American Psychologist, 45,* 1319-1324.

McCubbin, H. I., & Thompson, A. I. (1987). FTS: Family Traditions Scale. In H. I. McCubbin & A. I. Thompson (Eds.)., *Family assessment inventories for research and practice* (pp. 163-165). Madison: University of Wisconsin-Madison, Family Stress Coping ad Health Project.

McCubbin, H. I, McCubbin, M. A., & Thompson, A. I. (1987). FTRI. Family time and Routines Index. In H. I. McCubbin & A. I. Thompson (Eds.)., *Family assessment inventories for research and practice* (pp. 133-141). Madison: University of Wisconsin-Madison, Family Stress Coping ad Health Project.

McIntosh, R. P. (1985). *The background of ecology: Concepts and theory.* Cambridge: Cambridge University Press.

McLloyd, V. (1999). The impact of economic hardship on black families and children: Psychological distress, parenting, and socio emotional development. *Child Development, 61*, 311-346.

McLuhan, M. & Fiore, Q. (1967). *The medium is the message.* New York: Random House.

McLuhan, M.,& Powers, B. R. (1989). *The global village: Transformations in world life and media in the 21st century.* New York: Oxford University Press.

Melson, G. (1980). *Family and environment: An ecosystem perspective.* Minneapolis, Minnesota: Burgess Publishing Company

Merchant, C. (1992). *Radical ecology. The search for a livable world.* New York, London.

Micklin, M. (1973). Introduction: A framework for the study of human ecology. In M. Micklin (Ed.), *Population, environment and social organization* (pp. 2-19). Hillsdale, Illinois: Dryden.

Micklin, M. (1984). The ecological perspective in the social sciences: A comparative overview. In M. Micklin & H. Chaladin (Eds.). *Sociological human ecology: Contemporary issues and applications.* (pp. 51-90). Boulder: Westview Press.

Milgram, S. (1970). The experience of living in cities. *Science, 167*, 1461, 1468.

Miller, J. (1978). *Living systems.* New York: McGraw-Hill.

Naess, A. (1973). The shallow and the deep, long-range ecology movements: A summary. *Inquiry, 16*, 95-100.

Naess, A. (1989). *Ecology, community and lifestyle.* Translated and revised by D. Rothenberg. Cambridge: Cambridge University Press.

Naess, A. (1993a). The deep ecology movement: Some philosophical aspects. In M.E. Zimmerman, J.B. Callicott, G. Sessions, K.J. Warren and J. Clark, (Eds.) *Environmental philosophy.* (pp. 193-212). Englewood Cliffs, NJ: Prentice Hall.

Naess, A. (1993b). Simple in means, rich in ends. In M.E. Zimmerman, J.B. Callicott, G. Sessions, K.J. Warren and J. Clark, (Eds.) *Environmental philosophy.* (pp. 182-192). Englewood Cliffs, NJ.: Prentice-Hall.

Naar, J. (1990). *Design for a livable planet.* New York: Harper and Row.

Neff, J. (1983). Urbanicity and depression reconsidered. The evidence regarding depressive symptomatology. *The Journal of Nervous and Mental Disease, 171,* 546-552.

Neisser, U. (1967). *Cognitive psychology.* New York: Appleton-Century-Crofts.

Odum, H. T. (1994). *Ecological and general systems: An introduction to systems ecology.* Revised edition. Niwot, Colorado: The University Press of Colorado.

Odum, E. P. (1989) *Ecology and our endangered life-support systems.* Sunderland, MA: Sinauer Associates.

Odum, H. ,& Odum, E. (1976). *Energy basis for man and nature.* New York: McGraw-Hill.

Ooi, G. L. & Kwok, K.(Eds.)(1997). *City and the state: Singapore's built environment.* Oxford University Press.

Orlove, B. S. (1980). Ecological anthropology. *Annual Review of Anthropology, 9,* 236-273.

Orr, D. (1992). *Ecological literacy.* Albany: State University of New York Press.

Oskamp, S. (2000). A sustainable future for humanity. *American Psychologist, 55,* 496-508.

Oyemade, U. (1988). An ecological study of factors affecting behavioral outcomes of adolescents. In J. Borden & J. Jacobs (Eds.) *Human ecology: Research and applications* .(pp. 146-153). College Park, Maryland: Society for Human Ecology.

Paolucci, B., Hall, O., & Axinn, N. (1977). *Family decision making: An ecosystem approach.* New York: John Wiley & Sons.

Park, R. E. (1915). The city: Suggestions for the investigation of human behavior in the city environment. *American Journal of Sociology, 20,* 577-612.

Park, R. E. (1923). The natural history of the newspaper. *American Journal of Sociology, 29,* 273-289.

Pert, C. B. (1997). *The molecules of emotion.* New York: Scribner.

Piaget, J. (1950). *The psychology of intelligence.* London: Routledge & Kegan Paul.

Piaget, J. (1963). *The origins of intelligence in children* (2nd ed.). New York: Norton.

Piaget, J. (1972). *Psychology and epistemology*. London: Penguin.

Piaget, J. (1973). *The child and reality*. New York: Viking Press.

Platt, R. H., Rowntree, R. A. & Muick, P. C. (Eds.)(1994). *The ecological city. Preserving and restoring urban biodiversity*. Amherst, Massachusetts: University of Massachusetts Press.

Plous, S. (1993). *The psychology of judgment and decision making*. New York: McGraw-Hill.

Raffestin, C., & Lawrence, R. (1990). An ecological perspective on housing, health, and well-being. *Journal of Sociology and Social Welfare, 17*, 143-160.

Rainwater, L. (1966). Fear and house-as-haven in the lower class. *Journal of the American Institute of Planners, 32*, 23-31.

Regan, T. (1981). The nature and possibility of an environmental ethic. *Environmental Ethics, 3*, 181.

Roehlkepartain, J. L. (1997). *Building assets together: 135 group activities for helping youth succeed*. Rev. ed. Minneapolis, Minn.: Search Institute.

Rogers, C. R. (1961). *On becoming a person: A therapist's view of psychotherapy*. Boston: Houghton Mifflin.

Rogers, C. R. (1980). *A way of being*, Boston: Houghton Mifflin.

Rokeach, M. (1973). *The nature of human values*. New York: Free Press.

Roseland, M. (Ed.)(1997). *Eco-city dimensions: Healthy communities, healthy planet*. Gabriola Island, B.C.: New Society Publishers.

Rosenau, P. M. (1992). *Postmodernism and the social sciences*. Princeton, NJ: Princeton University Press.

Roszak, T., Gomes, M. E., & Kanner, A. D. (Eds.)(1995). *Ecopsychology*. San Francisco: Sierra Club Books.

Roy, R. (1979). Interdisciplinary science on campus: The elusive dream. In J. Kockelmans (Ed.) *Interdisciplinarity and higher education.*(pp. 161-196). University Park: The Pennsylvania State University Press.

Rutter, M. (1981). The city and the child. *American Journal of Orthopsychiatry, 51,* 610-625.

Safilios-Rothschild, C. (1970). The study of family power structure: A review 1960-1969. *Journal of Marriage and the Family, 32 ,*539-551.

Sampson, E. E. (1983). Deconstructing psychology's subject. *Journal of Mind and Behavior, 4,* 136-164.

Sampson, E. E. (1985). The decentralization of identity: Towards a revised concept of personal and social order. *American Psychologist, 40,* 1203-1211.

Sarup, M. (1993) *An introductory guide to post-structuralism and postmodernism.* Atlanta: University of Georgia Press.

Scales, P. C. (1999). Reducing risks and building developmental assets: Essential actions for promoting adolescent health. *The Journal of School Health, 69,* 113-19.

Scarr, S. (1992). Developmental theories for the 1990's: Development and individual differences. *Child Development, 63,* 1-19.

Scarr, S., & McCartney, K. (1983). How people make their own environments: A theory of genotype - environment effects. *Child Development, 54,* 424-435.

Schell, L. & Ulijaszek, S. J. (1999). *Urbanism, health and human biology in industrialised countries.* Society for the Study of Human Biology Symposium Series, Vol. 40. Cambridge University Press.

Schorr, L. (1989). *Within our reach.* New York: An Anchor Book - Doubleday Publishing Group.

Senge, P. M. (1990). *The fifth discipline.* New York: Doubleday.

Sessions, G. (1993). Introduction. In M.E. Zimmerman, J.B. Callicott, G. Sessions, K.J. Warren & J. Clark, (Eds.) *Environmental philosophy.* (pp. 161-170). Englewood Cliffs, N.J.: Prentice-Hall.

Simmel, G. (1917). *Grundfragen der Soziologie (Individuum und Gesellschaft) von Georg Simmel.* Berlin, Leipzig, G. J. Gööschen.

Simmel, G. (1971). *On individuality and social forms; selected writings.* Edited and with an introd. by Donald N. Levine. Chicago: University of Chicago Press.

Skinner, B.F. (1971). *Beyond freedom and dignity.* New York: Knopf.

Smith, R.(1972). *The ecology of man: An ecosystem approach.* New York: Harper and Row.

Smith, M. P. (1979). *The city and social theory.* New York: St Martin's Press.

Snyder, G. (1974). *Turtle Island.* New York: New Directions.

Stephan, G. E. (1970). The concept of community in human ecology. *The Pacific Sociological Review, 13,* 218-228.

Stern, P. C. (2000). Psychology and the science of human-environment interactions. *American Psychologist, 55,* 523-530.

Stewart, P. (1986). Meaning in human ecology. In R. J. Borden (Ed.) *Human ecology: A gathering of perspectives.* (pp. 109-116). College Park: The Society for Human Ecology.

Sylvan, R.(1985a). A critique of deep ecology. Part 1. *Radical Philosophy, 40,* 2-12.

Sylvan, R.(1985b). A critique of deep ecology. Part 2. *Radical Philosophy, 41,* 10-22.

Talbot, M. (1986). *Beyond the quantum.* New York: Macmillan.

Tansley, A. G. (1939). British ecology during the past quarter century: The plant community and the ecosystem. *Journal of Ecology, 27,* 513-530.

Thomashow, M.(1995). *Ecological identity: Becoming a reflective environmentalist.* Cambridge, Mass.: MIT Press.

Tonnies, F. (1940). *Fundamental concepts of sociology (Gemeinschaft und gesellschaft).* Translated and supplemented by C.P. Loomis. New York: American Book Company.

Visvader, J. (1986). Philosophy and human ecology. In R. J. Borden (Ed.) *Human ecology: A gathering of perspectives,* pp. 117-127. College Park: The Society for Human Ecology.

Vygotsky, L. S. (1962). *Thought and language.* Cambridge, MA: MIT Press.

Vygotsky, L. S.(1963). Learning and mental development atr school age. In B. Simon & J. Simon (Eds.), *Educational Psychology in the USSR.* (pp. 21-34). London: Routledge & Kegan Paul.

Wargo, J. (1998). *Our children's toxic legacy.* New Haven: Yale University Press.

Warren, R. (1998). *The urban oasis: Guideways and greenways in the human environment.* New York: McGraw-Hill.

Werner E. E., & Smith, R. S. (1992). *Overcoming the odds: High risk children from birth to adulthood.* Ithaca, NY: Cornell University Press.

Westney, O.E., Brabble, E. W., Saxton, C., & Holloman, L. (1986). Concepts of human ecology. In C.E. Edwards (Ed.) *Human Ecology Monograph*, (10th ed.). Washington, D.C.: Howard University.

Whyte, W. F. (1980). *The social life of small urban spaces.* Washington, D.C.: The Conservation Foundation.

Wiener, N. (1948). *Cybernetics.* Cambridge: Technology Press.

Wiener, N. (1967). *The human use of human beings.* New York: Avon Books.

Wirth, L. (1938). Urbanism as a way of life. *American Journal of Sociology, 44,* 1-24.

Winter, D. D. (1996). *Ecological psychology: Healing the split between planet and self.* New York: Harper Collins.

Winter, D. D. (2000). Some big ideas for some big problems. *American Psychologist, 55,* 516-522.

Winthrop, H. (1964). Methodological and hermeneutic functions in interdisciplinary education. *Educational Theory, 14,* 118-127.

Yancey, W. L. (1971). Architecture, interaction and social control: The case of a large-scale public housing project. *Environment and Behavior, 3,* 3-21.

Young, G. L. & Broussard, C. A. (1986). The species problem in human ecology. In R. J. Borden (Ed.) *Human ecology: A gathering of perspectives.* (pp. 55-67).College Park, Maryland: The Society for Human Ecology.

Zimbardo, P. G. (1973). The human choice: Individuation reason, and order vs deindividuation, impulse and chaos. In J. Helmer and N. A. Eddington (Eds.) *Urbanman: The psychology of urban survival.* (pp. 196-238). New York: The Free Press

Zimmerman, M. E. (1987): Feminism, deep ecology, and environmental ethics. *Environmental Ethics, 9*, 21-44.

Zimmerman, M. E. (1993). Rethinking the Heidegger - deep ecology relationship. *Environmental Ethics, 15*, 195-224.

Index

acting system 20
adaptation 29, 50, 55, 68
affect 46, 52
American Home Economics
               Association 40
analytical boundaries 24
anima 66
animus 66
artifacts 42
assets 76
atomistic families 48
authority 52

behavior change 120
behavior settings 69
behavioral production 67
bidirectional time 55
bio-social system 44
bioecological model 75
biological ecology 9, 110
bisociation 10
body territories 56
boundaries 24
Bronfenbrenner's model 69
Butterfly Effect 111
bystander apathy 87

capitalism 93
change 29
Chicago School of urban sociology 86
choices 55
chronosystem 71
closed family type 48
closed systems 18
cohort 71
collective unconscious 66
communication 28, 51
communication skills 53
community 9
community ecology 8

complexity 21, 25, 42
component 25
conceptual structure 10
conflict 29
conscious 65
consciousness 68
continuation 32
control 28
controlled systems 19
conversion 23
critical science 98
crowding 88
cultural ecology 7
culture 5
culture of poverty 46
cybernetic control 23
cybernetic system 19
cybernetics 28, 41

decay 29, 32
decision making 22, 52
deconstruction 122
deep ecology 7, 109, 113, 114, 116
defensive moves 57
description 6
developmentally appropriate 74
dialectical 12
dialectical naturalism 94
domination 93, 115
dynamic 32
dynamic components 26
dynamic equilibrium 22, 28
dynamic system 23

E-E 40
Earth 109
ecofeminists 115
ecological constellations 86
ecological method 7
ecological synthesis 31

ecological systems	21
ecological systems theory	69
ecological units	86
ecology	39
ecosophy	113
ecosystem	7, 9, 39
emancipatory	94
emergence	17, 32
energy	22, 27, 46-48, 55, 79
energy laws	27
enmeshed family	48
enmeshed family system	45
entropy (disorder)	29
environment	21, 41
environmental justice	95, 96
environmental policy	89
environmental psychology	69
environmental quality	89
environmental racism	95
environmental systems	21
epidemiological risk models	76
equilibrium	22
evolution	29
exchange	54
exosystem	44, 70, 73
explanation	6
familism	45
family	42
family ecological perspective	40
family ecology	39
family ecosystem	57
family relations	41
family resource management	41
family traditions	56
family values	22
feedback	22, 28
fueling	47
functional boundaries	24
future orienting	55
Gaia Hypothesis	112
Gemeinschaft	85
general systems	11, 20
generational influences	71
genogram	55
Gesellschaft	85
global village	85
goals	28, 42, 52
governing components	27
Green Party	96
growth	29
habitat	118
hierarchies of power	122
hierarchy	25, 42, 93
high quality energy	47
holism	6, 29
holistic	30, 32
holistic ecological analysis	89
holistic thinking	11
home economics	40
home oekology	40
home territories	56
homeostasis	28
homeostatic system	23
human constructed environment	4
human development	41
human domination	115
human ecology	21, 57, 67, 87
human ecosystems	93
human lifestyles	120
human resources	49
human-Earth relationships	113
humanistic Views	67
idiographic science	122
immediate closure	53
independence	29
indigenous populations	95
individual differences	75
information	28, 41
information overload	88
information processing components	28
input	22
interaction	29

interactional territories	56
interconnectedness	11
interdependence	29
interdisciplinary	6
interface	25
internal environments	65
internal flows	27
internal system	42
interpersonal subsystem	43
interrelationship components	28
interruptive components	29
intuitive thinking	53
irrationalities	122
kinetic energy	27, 79
laboratory controls	32
land use	87
lead	4
learning	28
level 1 feedback	23
level 2 feedback	23
level 3 feedback	23
level 4 feedback	23
levels of complexity	23
life process components	29
life space	69
linear change	105
linear time	55
living systems	79
macrosystem	45, 70
macrotime	71
market society	94
materials	41
matter	27, 55
meaning	46, 52
mechanistic thinking	53
mental health	88
mercury	4
mesosystem	70, 78
mesotime	71
Michigan State University	41
microsystem	44, 70
microsystems	72, 78
microtime	71
modernity	105
money	49
morphogenesis	23
morphogenic systems	19
morphostatic systems	19
multidimensional	5
multifinality	3
mutual interaction	29
natural ecosystems	93
natural environment	41, 57
needs	42, 44
negative entropy	27
negative feedback	22
new age	68
new paradigm	1
new physics	111
new urbanism	88
niche	8
non-collateral systems	43
non-temporal orienting	55
nonrenewable resources	27
nurturance	52
O-O	40
objective elimination	53
objective world	68
oekology	39, 40
offensive moves	57
open systems	18, 27, 48
organism	9
organization	55, 68
outputs	21
oversimplifying	2
paradigm	31
paradigm level	31
participatory	94
past orienting	55
patterned systems	20
patterns of stability	56

perception	50
person-environment interactions	77
personal subsystem	43
physical well being	56
physics of energy	27
physiological boundary	24
political ecology	7
population density	88
positive feedback	22
postmodernism	105, 106
postmodernity	105
potential energy	41, 47
power	46, 52
preconscious	65
preference ranking	53
present orienting	55
privacy	57
process level	31
progress model	106
proximal process	76, 78
psychoanalysis	65
psychological ecology	69, 77, 79, 120
psychological ego	113
psychological environments	65, 67
psychological overload	88
psychological well being	56
psychosocial approach	66
public territories	56
purposes	28
quality of life	8
quantum theory	111
radical ecology	96, 97
random family system	45
rational decisions	94
re-orientation	23
reductionist	8
regime level	31
reproduction	29
resource channels	46
resources	42
role	43
role	43
rules	22, 28
schema	68
school desegregation	97
Search Institute	76
seasonal affective disorder	4
self-actualization	67
self-efficacy	67
shallow ecology	110
simple feedback	23
Skeptical postmodernists	106
social construction	118
social deviancy	88
social ecology	7, 93, 94
social environments	4
social justice	97
social-emotional energy	47
socio-cultural environment	41
space	46
species	8
spouse subsystem	43
strain	29
stress	29
structural analysis	77
structural changes	29
structural components	25
structural model	77
structure	25
subsystem	32, 43
succession	87
suprasystem	32
sustainability	89, 109, 119
sustainable environments	123
sustenance activities	54
symbiosis	94
symbolic representations	67
systems	17, 25, 30, 32
systems thinking	53
technological advances	120
termination	29, 30

threat 29
throughput 22
time 46, 55
transactions 42
transdisciplinary 6, 10
triadic determinism 66
types of systems 17

unconscious 65, 66
uncontrolled systems 19
unidirectional time 55
urban centers 86
urban cosmology 89
urban ecology 85
urban laboratories 87
urban sustainability 90
urban vegetation 89
urban way of life 86

value orientation 46
value structures 120
values 8, 42, 45
verbal language 30
voluntary simplicity 120
vulnerability 76

wildlife ecology 110